Applied Probability

MATHEMATICAL CONCEPTS AND METHODS
IN SCIENCE AND ENGINEERING

Series Editor: **Angelo Miele**
  Mechanical Engineering and Mathematical Sciences, Rice University

Applied Probability

Frank A. Haight

The Pennsylvania State University
University Park, Pennsylvania

PLENUM PRESS · NEW YORK AND LONDON

Library of Congress Cataloging in Publication Data

Haight, Frank A.
    Applied probability.

    (Mathematical concepts and methods in science and engineering–; 23)
    Bibliography: p.
    Includes index.
    1. Probabilities. I. Title. II. Series.
QA273.H317   519.2   81-4690
ISBN 978-1-4615-6469-0   ISBN 978-1-4615-6467-6 (eBook)   AACR2
DOI 10.1007/978-1-4615-6467-6

©1981 Plenum Press, New York
Softcover reprint of the hardcover 1st edition 1981
A Division of Plenum Publishing Corporation
233 Spring Street, New York, N.Y. 10013

For my dear children, Julian and Molly

Preface

Probability (including stochastic processes) is now being applied to virtually every academic discipline, especially to the sciences. An area of substantial application is that known as operations research or industrial engineering, which incorporates subjects such as queueing theory, optimization, and network flow.

This book provides a compact introduction to that field for students with minimal preparation, knowing mainly calculus and having "mathematical maturity." Beginning with the basics of probability, the development is self-contained but not abstract, that is, without measure theory and its probabilistic counterpart.

Although the text is reasonably short, a course based on this book will normally occupy two semesters or three quarters. There are many points in the discussions and problems which require the assistance of an instructor for completeness and clarity.

The book is designed to give equal emphasis to those applications which motivate the subject and to appropriate mathematical techniques. Thus, the student who has successfully completed the course is ready to turn in either of two directions: towards direct study of research papers in operations research, or towards a course in abstract probability, for which this text provides the intuitive background.

Pennsylvania State University Frank A. Haight

Contents

1

Discrete Probability

1.1. Applied Probability

Applied mathematics deals with the quantitative description of phenomena. A formulation in mathematical terms is called a "mathematical model" of a phenomenon. Thus a parabola models a thrown stone, an ellipse, the orbit of a planet. It is an axiom of science that measurements made under identical conditions will give identical results, and therefore if the conditions under which the phenomenon is observed are fixed, the mathematical model is determined. Measurements of length, temperature, and voltage are typical deterministic experiments.

Some phenomena, however, can give different results when repeated observations are performed. A simple case is the toss of a coin. The fact that the coin will fall to earth is deterministic, but the observation of "heads" or "tails" seems to violate the basic axiom of science. How can this be understood? A particle counter which records cosmic rays may give a different number of registrations during equal time periods. Is this consistent with scientific determinism? The fact is that the conditions of observation are not identical. Subtle differences in the dynamics of the coin toss account for the variable results. With a substantial investment of time and money, a predictable coin-tossing machine could no doubt be invented. The uncertainty arises specifically from our ignorance of the scientific conditions governing the toss. The same must be true of the cosmic-ray recorder, although a deterministic prediction of the number of incoming particles is clearly impossible. The number can be regarded as "unknowable." To a greater or lesser degree, this ignorance of outcome makes a phenomenon appear to be probabilistic rather than deterministic.

The distinction between probabilistic and deterministic phenomena is useful, and necessary in building mathematical models, but it is still fundamentally pragmatic. There are gray areas. The simplest example of gradual transition from determinism to probability is in the measurement of length. If the calibration of the ruler is gross, the same result will be obtained each time, and the length will be modeled by a real, positive number. As the calibration becomes finer and finer, the chance of "error" in measurement increases and the mathematical model for the length will be not a single number, but a range of values together with a probability measure specifying the likelihood of each value's occurrence.

Thus, in modeling, it is first necessary to decide whether the evidence of the phenomenon favors a deterministic or probabilistic model. Where ignorance of the exact conditions of the experiment is somewhat superficial, either course might be appropriate, but where ignorance is truly abysmal, as in the cosmic-ray example, probability offers the only satisfactory descriptive system.

This book is concerned with the construction and analysis of probability models, with examples from real problems of operations research and especially queueing theory. By studying the principles given and their application, the student will gradually learn to build models for new problems that may be encountered in science, engineering, management, and industry. This book does not deal with the statistical analysis of data obtained from probabilistic experiments; rather, it forms the basis for statistics.[†]

A probability model contains three elements: a statement of what can possibly happen (sample space), a statement of the probabilities, and a law attaching the probabilities to the elements of the sample space. It is possible in theoretical probability to construct these elements in a mathematically exact manner; the result is comparable to other branches of pure mathematics. In this book, however, the models are built on the basis of example and intuition. In this way the book can equally well be used as an introduction to theoretical probability, since intuitive understanding is the best basis for exact understanding.

The elements of a probability model will be discussed in the following sections.

[†]A statistical sample is regarded as a sequence of n "independent" random variables; the emphasis in this book is on two or three not necessarily independent random variables.

1.2. Sample Spaces[†]

The description of a probabilistic experiment begins with a simple list of all the things that can possibly happen, whether they are likely or not. This is not always quite as simple as it might appear. In the coin-tossing experiment, is it right to allow for the possibility of the coin standing on edge? This (and other exotic possibilities) depends on the *definition* of the experiment; according to logic, the purpose of the model, and the needs of the modeler, it might be defined either way. Conventionally, only two possibilities are allowed: heads and tails. This does not reflect any conviction about the actual nature of real-life coins. On the contrary, the "coin" is only a convenient description of the simplest kind of experiment in which only two outcomes are possible. If the outcomes are postulated to be equally likely, the expression is a "fair coin." If for some other reason it would be useful to discuss an experiment with two rather high probability outcomes and one very low probability outcome, the image of a sample space consisting of *head*, *tail*, and *edge* might be employed.

In the examples which follow, it will be useful for the student to think about the sample spaces given in comparison with various alternatives.

Example 1. Five repairmen service a large number of machines, each of which may or may not need repair at a given moment. Sample space: the number of repairmen busy, $0, 1, 2, 3, 4, 5$.

Example 2. Two chess players agree to a match where the winner will be the first one to accumulate five points, with one point awarded for a win, half a point for a draw, and no points for a loss. Sample space: the number of games played in the match, $5, 6, 7, 8, 9, 10$. (Note: If 10, the match is a draw.)

Example 3. A mass-produced item is checked by five checkers and passed if at least four of the checkers pass it. Sample space: the number of checkers rejecting a passed item, $0, 1$.

[†]The term "sample space" for the possible outcomes of a probabilistic experiment has some drawbacks: In the first place it could suggest statistical sampling, and in the second place it is more akin to the "domain" of a variable, as that term is used in mathematics. Nevertheless, it is a well-established expression and is used here as in virtually all textbooks.

Example 4. The capacity of an elevator is five persons. Sample space: the number of persons in the elevator at a given moment, $0, 1, 2, 3, 4, 5$.

Example 5. m salesmen are assigned to n territories, with $m > n$. Sample space: the number of territories without a salesman, $0, 1, \ldots, n-1$.

Example 6. m salesmen are assigned to n territories, with $m < n$. Sample space: the number of territories without a salesman, $n-m, n-m+1, \ldots, n-1$.

In the first six examples, the number of outcomes in the sample space is finite. The outcomes are called *points* of the sample space. In the examples which follow, the sample spaces contain an infinite number of points.

Example 7. A repairman attends to a large number of machines. If a machine is broken when he finishes a job, he goes directly to the new job. Sample space: the number of jobs completed between periods of idleness, $1, 2, 3, 4, \ldots$. The possibility may also exist that once a repair is started, he may never again (in theory) be idle. To account for this possibility it may also be necessary to define a point in the sample space by a symbol denoting "never again idle."

Example 8. Light bulbs are tested as they come off the production line and are found to be working or defective. Sample space: the number of bulbs tested before the first defective is found, $0, 1, 2, \ldots$. Should the possibility of an infinite duration of good bulbs be built into this model, similar to Example 7? This is a typical problem in model building, and the answer will depend on the nature of the model needed.

Example 9. There are n keys, only one of which fits a lock, and keys are chosen at random† and tried, not eliminating those which have failed. Sample space: the number of keys tried before a success, $1, 2, 3, 4, \ldots$. Another sample space would be the number of failures before a success, $0, 1, 2, 3, \ldots$.

Example 10. Bids are invited for a contract. Sample space: the number of bidders, $0, 1, 2, 3, \ldots$. In this case it is difficult, if not impossible, to

†For the time being, the expression "at random" is used in a purely intuitive sense; later it will be made more precise by specifying equal probabilities.

imagine the need for an infinite number of bidders. Such an assumption at least made logical sense in Examples 7, 8, and 9, but it is not at all meaningful to imagine an infinite number of bidders.

Example 11. The number of cosmic particles registered by a counter in a finite time. Sample space: $0, 1, 2, 3, \ldots$, provided the counter mechanism permits registrations arbitrarily near one another.

Example 12. The number of bus tickets bought by a single customer. Sample space, $1, 2, 3, 4, \ldots$ Is there a theoretical maximum?

Comparing these examples with one another, it seems that the choice of the initial value is usually not so difficult. In Example 10, there could presumably be zero bidders for the contract, so it would be inappropriate to begin with one. On the other hand, in Example 12, the idea of buying zero bus tickets makes no sense, so the sample space properly begins with one. In Example 4, the sample space might include the point zero for a self-service elevator, but if an operator is present, the sample space might begin with one.

Although a clear understanding of the system being modeled often gives the initial value of the sample space, genuine ambiguities are sometimes encountered in deciding on the final value. In Example 12, it certainly does not make sense to assume an enormously large number of people buying a ticket, but on the other hand, where should the line be drawn? Before going further into this question, it is useful to consider some further examples where the problem is even more difficult.

Example 13. Consider the number of animals of a certain species born in a litter. Sample space: $1, 2, 3, \ldots$ Should there be a maximum value?

Example 14. Consider the number of points scored in a game. Sample space: $0, 1, 2, \ldots$, provided the kind of game is not specified. Are there games where certain values would need to be omitted? Does a maximum value make sense?

Example 15. Consider the number of cars observed in a block. Sample space: $0, 1, 2, 3, \ldots$ It would presumably be possible to calculate a maximum possible value.

Example 16. Consider the number of insects of a certain type found on a leaf. Sample space: $0, 1, 2, \ldots$. Is there any logical way of specifying a maximum value?

Example 17. Consider the number of misprints on a page of a book. Sample space: $0, 1, 2, 3, \ldots$. The number surely must be rather small, but it is difficult to specify any exact maximum.

Example 18. Consider the number of letters mailed at once. Sample space: $1, 2, 3, \ldots$.

In examples of this type, it is hard to know whether to assume a finite or infinite sample space. In general, there are two disadvantages to the choice of a finite domain: not knowing which value to choose for the maximum, and the fact that the mathematical formulation is usually more difficult.

The latter point may seem surprising at first, but a moment's thought may give a clue as to why it is often easier to deal with infinite sample spaces: Sums of infinite series are often simpler than the corresponding "partial sums" stopping at an arbitrary finite value.

The principal argument against a sample space which is open ended is that very large values (such as 10^{10}) are either biologically impossible (as in Example 13) or difficult to imagine (as in Example 16) or so unlikely as to be absurd (as in Examples 14 and 18). On the other hand, once the probabilities are defined, and assigned to the points of the sample space, it may (and indeed should!) turn out that these enormous, absurd values will correspond to very small probabilities (say, $10^{-10^{10}}$). In fact, it almost always turns out in practice that arguments about "realism" do not lead to substantially different models, while the mathematical convenience of an infinite sample space may be considerable.

It is worth emphasizing that the sample space does not come automatically from the definition of the experiment. For example, if a car is selected, one might look at the number of passengers, the number of cylinders, the number of owners, or the number of parking tickets.

Also, a sample space does not need to consist of numbers; for a car, one might look at the color or the make. This point will be discussed more fully in Section 1.4.

Finally, the sample spaces given in this section are all discrete, whether finite or infinite. Beginning in Chapter 4, some continuous sample spaces will be treated. Many students are not as familiar with the mathematical techniques of discrete variables (especially summation) as with those for continuous variables (integration), and so in this first chapter, some emphasis is given to technique, while the fundamentals of probability are being discussed. As the book progresses, more and more of the technique will be taken for granted, and the steps in proofs correspondingly abbreviated.

1.3. Probability Distributions and Parameters

The second ingredient in a probability model is a set of "probabilities," that is, numbers which can be used to measure the probabilities of various points in a sample space. By definition, a set of numbers, each of which is between zero and one, such that the entire set sums to unity is a set of probabilities, or *probability distribution*. (The idea of the word "distribution" is that the total available probability is distributed among the various points of a sample space. The definition of a probability distribution corresponds rather closely to common speech, except that people sometimes speak of 100% for certainty, rather than $+1$.)

Thus the set of numbers $(\frac{1}{2}, \frac{1}{2})$ constitutes a valid probability distribution, as does $(\frac{1}{2}, \frac{1}{4}, \frac{1}{4})$. However, $(\frac{1}{2}, \frac{1}{2}, \frac{1}{2})$ is not a probability distribution, since the total exceeds unity, and $(-\frac{1}{2}, 1, \frac{1}{2})$ is not, even though the total is one, because one of the values does not lie in the unit interval between zero and one.

The definition of a probability distribution is a loose one, in that there are any number of sets of numbers that can satisfy it. This being the case, it is rather remarkable that the number of different distributions actually used in practice is rather small. In this section a few of the most basic distributions will be introduced.

First, consider the fact that any finite set of positive numbers can be used to form a probability distribution by dividing by their sum. Thus the set $(3, 7, 17)$ leads to the valid probability distribution $(\frac{3}{27}, \frac{7}{27}, \frac{17}{27})$. In algebraic terms this can be expressed by saying that any terms forming a (finite or infinite) series can be converted into a probability distribution by dividing each term by the sum of the series, provided only that the terms be positive. This process of converting positive terms into a probability distribution is called *normalizing to unity*; it is a technique often used.

Second, remember that the probabilities in a distribution do not need to be arithmetic, but can be symbols. Thus

$$(p, 1-p), \qquad 0 < p < 1,$$

is a probability distribution for values of p indicated by the inequality. Where a letter occurs in the distribution, it is called a *parameter* of the distribution. It usually happens that the expressions are acceptable as probabilities only for a certain range of parameter values, as in the example above. Therefore it is essential in writing a probability distribution which involves a parameter to specify the allowable range of parameter values. Certainly $(p, 1-p)$ is not a probability distribution for $p=2$ or $p=-\frac{1}{2}$.

The Rectangular Distribution. When all the probabilities in a distribution are equal, the distribution is called rectangular. Examples of a rectangular distribution are $(\frac{1}{2}, \frac{1}{2})$ and $(\frac{1}{6}, \frac{1}{6}, \frac{1}{6}, \frac{1}{6}, \frac{1}{6}, \frac{1}{6})$. In general, a rectangular distribution is represented by a list of n values of $1/n$, where n, the parameter in the distribution, must be a positive integer.

The Poisson Distribution. The Poisson[†] distribution is formed by normalizing the terms in the exponential sum to unity:

$$e^x = 1 + x + \frac{1}{2} x^2 + \frac{1}{3!} x^3 + \cdots .$$

The parameter is usually represented by the letter λ; the probabilities are therefore written

$$\left(e^{-\lambda}, \lambda e^{-\lambda}, \frac{1}{2} \lambda^2 e^{-\lambda}, \frac{1}{3!} \lambda^3 e^{-\lambda}, \ldots \right),$$

where $\lambda > 0$. Obviously this set of probabilities is going to be applicable to infinite sample spaces, since there are infinitely many terms in the series.

It is usually convenient to represent all the terms in a probability distribution by a single formula, with a "dummy" variable introduced for compact notation. For example, the Poisson distribution can be written

$$\frac{\lambda^x e^{-\lambda}}{x!}, \qquad x = 0, 1, 2, \ldots, \lambda > 0. \tag{1}$$

[†]S. D. Poisson, French mathematician, 1781–1840.

Written in this way, there is often an easy and natural correspondence between the values of the dummy variable x and points of some sample spaces such as those discussed in Section 1.2.

   The Geometric Distribution. If the geometric series

$$1+\rho+\rho^2+\rho^3+\cdots=\frac{1}{1-\rho}$$

is normalized to unity, and the resulting probabilities written in compact form, the geometric distribution is obtained:

$$(1-\rho)\rho^x, \qquad x=0,1,2,\ldots, \ 0<\rho<1. \tag{2}$$

This distribution has a parameter on the unit interval and is defined over an infinite number of values.

   The Binomial Distribution. Sometimes the normalization to unity is obtained not by dividing the sum of the series, but by assigning parameter values. As an example, consider the binomial series

$$(\sigma+\pi)^n=\sum_{j=0}^{n}\binom{n}{j}\pi^j\sigma^{n-j}.$$

This well-known formula contains three parameters: σ and π, which may be any real numbers, and n, which is a positive integer (or possibly zero). By setting $\sigma=1-\pi$, the series is normalized to unity, and its terms can be used as probabilities:

$$\binom{n}{x}\pi^x(1-\pi)^{n-x}, \qquad x=0,1,\ldots, n, \ 0<\pi<1, n=0,1,2,\ldots. \tag{3}$$

This formula specifies a finite number $(n+1)$ of probabilities. Note that one of the two parameters of the distribution tells how many probabilities there are, while the other lies in the unit interval.

   The Fisher Distribution. The Fisher[†] distribution is obtained by normalizing the terms of the logarithmic series to unity and making other small changes:

$$\log(1+\beta)=\beta-\tfrac{1}{2}\beta^2+\tfrac{1}{3}\beta^3-\tfrac{1}{4}\beta^4+\cdots.$$

[†] R. A. Fisher, English statistician, 1890–1962.

The terms are alternating in sign and so cannot be used directly as probabilities. However, by replacing β by $-\beta$, each term is made negative. Then the whole expression is prefixed by a negative sign and normalized by division, giving the probabilities

$$\frac{-\beta^x}{x\log(1-\beta)}, \qquad x=1,2,3,\ldots, \quad -1\le\beta<1.$$

The five distributions introduced in this section will not provide for all the probability models required (new distributions occur from time to time in the sequel). They are given here to show how probability distributions can be formed. It will be useful for the student to write out some of the terms in each of these distributions, with specific parameter values, and see how they might be meaningful representations of probabilities for some of the examples given in Section 1.2. Here are some possibilities for investigation:

Rectangular distribution: Examples 3, 4
Poisson distribution: Examples 10, 11, 15, 16
Geometric distribution: Examples 8, 14, 17
Binomial distribution: Examples 1–6
Fisher distribution: Examples 7, 9, 12, 13, 18

Why is Example 3 listed twice?

These examples are given for practice with sample spaces and probability distributions, and there are not necessarily realistic connections between the two. That problem will be treated in the following section.

1.4. The Connection between Distributions and Sample Points: Random Variables

How is it possible to establish valid probabilities to attach to the points of a sample space? Consider first the two statements: (A) In tossing a fair coin, the probability of a heads is one-half, and (B) in tossing three fair coins, the probability that all three will come up heads is one-eighth.

Statement A might be written in symbolic form $P(\text{heads})=\frac{1}{2}$, or even $P(H)=\frac{1}{2}$, provided it is understood that we are referring to the toss of a fair coin. The equality sign links the right side of the equation, which contains a probability [in fact, one element of the probability distribution $(\frac{1}{2}, \frac{1}{2})$], with the left side, which contains a sample point enclosed by the "operator" $P(\ )$. The equality comes, as mentioned in Section 1.2, purely as a *definition* (of a "fair" coin). Statement B, on the other hand, is a *theorem* and needs to be proved.[†] In either case, the basic format is clear, and is of the form

$$P(\text{sample point}) = \text{probability},$$

and it is necessary to be sure that if all the sample points are listed with their corresponding probabilities, the total of the probabilities is unity. This means only that all of the sample points are accounted for and all the probabilities are used, with no leftovers on either side.

At this stage it is important to make the distinction between those sample points which are quantitative (represented by a number, such as the number of idle operators, the number of bugs on a leaf, etc.), on the one hand, and those which are qualitative (side of coin, color of car, name of individual, i.e., represented by a "label").

The qualitative case is useful in the beginning for introducing some elementary examples, and, in theory, for showing that not all probabilistic experiments must produce numerical outcomes. However, as the book progresses and more practical situations are under discussion, the experiments will be almost exclusively numerical. When the sample space is quantitative, we speak of a *random variable* (usually denoted by a capital letter). For example, let $X =$ the number of bugs on a leaf. The sample space, in this case $0, 1, 2, \ldots$, then represents the values that the random variable can take. Then the general displayed equation shown above takes on the special form

$$P(X = x) = \text{probability}, \qquad x = 0, 1, 2, \ldots.$$

There is often a natural correspondence between sample points and the values of the random variable, although in a completely theoretical sense the random variable is chosen, just as the sample space was chosen, as part of the definition of the experiment. In this chapter, random variable values will almost invariably be non-negative integers, or some subset (finite or infinite) thereof.[‡] There are many cases in which a random variable can be defined in a meaningful way for an experiment with purely qualitative sample points.

[†] There is a third category of probability statements encountered in the study of statistics: the empirical fact, established on the basis of experimental evidence.

[‡] The book as a whole treats positive (discrete or continuous) random variables.

In the toss of a coin, the labels "head" and "tail" can be replaced by the random variable $X=$ number of heads, with $X=0$ or $X=1$. Then the probabilities for a fair coin would be

$$P(X=0)=P(X=1)=\tfrac{1}{2}.$$

On the other hand, it would be difficult to assign meaning to a random variable in a qualitative experiment yielding, for example, colors or names. To say that $X=0$ means "RED," $X=1$ means "BLUE," and so forth would be logically permissible, but it would not shed much light on the situation. A good way of seeing the difference in these two examples is to anticipate Section 1.6 and notice that the concept of an average value makes sense for the random variable $X=$ number of heads; the average number of heads is $\tfrac{1}{2}$. The same meaning cannot be attached to colors.

It is not *always* possible, or when possible not always convenient, to compress the probability statements into a single formula. If, for example, the sample space consists of the values 0, 2, and 5, with probabilities

$$P(X=0)=\tfrac{1}{2},$$

$$P(X=2)=\tfrac{1}{4}-3^{1/2}/12,$$

$$P(X=5)=\tfrac{1}{4}+3^{1/2}/12,$$

it would hardly be worth the effort to try to reduce these three formulas to one comprehensive one.

Also, although the sample points and the values of the variable x often agree, as in the examples at the end of Section 1.3, it is not compulsory that they should do so. The probabilities can be assigned in any way at all. The Poisson probabilities fall naturally on the non-negative integers $0,1,2,3,\ldots$, but it is not difficult to assign them to values other than those. If the random variable X takes values which are multiples of 3, for example, $3,6,9,\ldots$, the Poisson assignment would be given by

$$P(X=3x)=\frac{\lambda^{x-1}e^{-\lambda}}{(x-1)!}, \qquad x=1,2,3,\ldots,$$

which is equivalent to

$$P(X=x)=\frac{\lambda^{x/3-1}e^{-\lambda}}{(x/3-1)!}, \qquad x=3,6,9,\ldots.$$

In other words, the sample points and the probabilities are completely separate entities which are combined to make a complete probability model.

Unfortunately, there is in the literature a considerable variability in notation, both because of different traditions and because of different purposes which need to be served. Often it is convenient to abbreviate $P(X=x)$ by p_x, representing the probability of the value x. If there is a parameter λ in the distribution, this might be included in the notation $p_x(\lambda)$, being, for example, the probability of the value x in a Poisson distribution with parameter λ. In some cases, mainly those which occur in theoretical probability, it is desirable to keep the random variable showing in the notation by writing $P(X=x)=p_X(x)$, but the right side is hardly more compact than the left side, and in this book the full form $P(X=x)$ will be retained when it is desirable to mention the random variable.

Omitting the random variable from the notation is convenient when it is fixed during an entire calculation and the calculation is rather complex. Examples of this will occur in Section 1.7 and subsequently.

Finally, it is worthwhile to comment on the theoretical nature of random variables. An abstract treatment of probability emphasizes that a random variable is a *function*, which takes sample points into numbers. That is, given any result of an experiment (sample point), the random variable translates this result into a number. This fact can be appreciated by considering different random variables for a single experiment. Suppose a coin is thrown. Let

X_1 = number of heads showing,

X_2 = time for the coin to come to rest,

X_3 = angle between the axis of the coin and floorboard cracks,

X_4 = distance of the center from the nearest wall,

X_5 = number of times the coin rotates in flight,

and so forth. The choice of one of these random variables is always a choice, and is not intrinsic in the experiment of throwing the coin. In other words, after the coin has been thrown, there is no *number* shining brilliantly from the coin to the observer; any number must be defined. In making the definition, the observer is choosing a function which must give a unique value for the result of any throw.

The same thing is true even in examples such as Example 10 of Section 1.2. There is no need to define $X =$ number of bids received. It could be the product of the ages of the bidders or the dollar value of the highest bid or the clock time when the first bid was received.

This book deals mainly with experiments involving easily defined random variables, but at a few critical points in the development, non-numerical sample spaces are employed for clarity. Also, it is henceforth assumed without further emphasis that random variables take non-negative values.

1.5. Events and Indicators

When a sample space is given and the corresponding probabilities are attached to each of the sample points, all information about the experimental probabilities is, in principle, known. If a random variable X is under consideration, this means that all expressions $P(X = x)$ are known. There may be other probabilities of interest, however: the probability that X is odd, that X is nonzero, that X is less than six, or that X takes the values 2 or 3.

Each such portion of the whole sample space is called an *event*; an event is a subset of the sample space. The original sample points are called *simple events* and events which contain more than one sample point are called *compound events*. It is often necessary to calculate the probability of compound events, and for this purpose the concepts and notation of set theory are useful.

Although some prior knowledge of these ideas will be helpful to the student, a very brief summary of notation as given below will permit the development of applied probability.

Consider two events E and F. All those sample points contained in either of the events are denoted by $E \cup F$, where the symbol $\cup$ can be read "or." Those points lying in both E and F are denoted by $E \cap F$, where $\cap$ is read "and." If every point of E is also a point of F, we write $E \subset F$ or, alternately, $F \supset E$. The symbols $\supset$ and $\subset$ can be read "includes" and "is included in"; they do not preclude the possibility that E and F are the same event. Some simple relationships between the symbols will be developed by example.

The Fundamental Principle of Probability. If E and F are two distinct simple events (i.e. sample points), then

$$P(E \cup F) = P(E) + P(F). \tag{4}$$

For random variables, this becomes

$$P(X=x \cup X=y) = P(X=x) + P(X=y), \qquad x \neq y. \tag{5}$$

In a theoretical treatment of probability this fact is given as an axiom.

Example 1. A fair die has six sides with the numbers 1, 2, 3, 4, 5, and 6 and a probability of one-sixth for each of the sides appearing in a single throw. Suppose a random variable X is defined as $X=$ number showing in one throw of a fair die, with possible values 1, 2, 3, 4, 5, and 6. Consider the event "X is odd." The probability can be calculated as follows:

$$P(X \text{ is odd}) = P(X=1 \cup X=3 \cup X=5)$$

$$= P(X=1) + P(X=3) + P(X=5)$$

$$= \tfrac{1}{6} + \tfrac{1}{6} + \tfrac{1}{6}$$

$$= \tfrac{1}{2}.$$

This example is trivial and the answer is obvious in any case. Now some less simplistic examples will be given, which will also permit a further acquaintance with probability distributions.

Example 2. The number of accidents per day in a factory has been found to have a geometric distribution with parameter ρ, where the parameter is estimated by statistical methods. If a random variable X is defined as the number of accidents in a given day, then

$$P(X=x) = (1-\rho)\rho^x, \qquad x=0,1,2,\ldots,\ 0<\rho<1.$$

What is the probability that there will be more than three accidents on a day?

$$P(X>3)=P(X=4\cup X=5\cup X=6\cup \cdots)$$

$$=P(X=4)+P(X=5)+P(X=6)+\cdots$$

$$=(1-\rho)\rho^4+(1-\rho)\rho^5+(1-\rho)\rho^6+\cdots$$

$$=(1-\rho)\rho^4(1+\rho+\rho^2+\cdots)$$

$$=\rho^4.$$

Example 3. It has been found that the number of claims received by an insurance company in a day is Poisson distributed with parameter λ, where the value of the parameter is obtained by statistical estimation. That is, if $X=$ number of claims received in a day, then

$$P(X=x)=\frac{\lambda^x e^{-\lambda}}{x!}, \qquad x=0,1,2,3,\ldots, \lambda>0.$$

What is the probability that a day will pass without more than one claim?

$$P(X\le 1)=P(X=0\cup X=1)$$

$$=P(X=0)+P(X=1)$$

$$=e^{-\lambda}+\lambda e^{-\lambda}$$

$$=(1+\lambda)e^{-\lambda}.$$

Example 4. There are n workers in an office and it is known that the number of workers reporting for work on a given day is binomially distributed with parameters n and π. In this case n is "fixed" by the terms of the problem and so could be considered more like a constant than a parameter. On the other hand, π would be estimated as in the earlier examples. What is the probability that there will be two or more absentees on a day selected at random? Let Y be the number of absentees and X, the number present for work, so that $X+Y=n$. Then, by hypothesis,

$$P(X=x)=\binom{n}{x}\pi^x(1-\pi)^{n-x}, \qquad x=0,1,\ldots, n, 0<\pi<1$$

and the probability of two or more absentees would be

$$P(Y \geq 2) = P(n - X \geq 2) = P(X \leq n - 2)$$

$$= P(X = 0 \cup X = 1 \cup X = 2 \cup \cdots \cup X = n - 2)$$

$$= P(X = 0) + P(X = 1) + \cdots P(X = n - 2)$$

$$= 1 - P(X = n - 1) - P(X = n)$$

$$= 1 - n\pi^{n-1}(1 - \pi) - \pi^n.$$

Probabilities for simple events have been assumed, according to the fundamental principle, to be additive. The same is true for compound events which are mutually exclusive, i.e., which have no sample points in common. When there are common points between E and F, additivity is no longer valid. In fact, if $P(E)$ and $P(F)$ are computed separately, all points in $E \cap F$ are counted twice, once for $P(E)$ and once for $P(F)$. Hence the additivity equation (4) needs to be modified as follows:

$$P(E \cup F) = P(E) + P(F) - P(E \cap F). \tag{6}$$

This formula is also an axiom of probability.

Example 5. In the throw of a single die, let $X =$ the number showing on top. Assume the die to be fair, and let $E = X$ is odd, and let $F = X$ is greater than 3. Then $P(E) = \frac{1}{2}$, $P(F) = \frac{1}{2}$, $P(E \cap F) = \frac{1}{6}$, and $P(E \cup F) = \frac{5}{6}$. These values satisfy Eq. (6).

Formulas (4) and (6) remain valid even if more than one random variable is involved.

Example 6. In an industrial quality-control experiment, five items are chosen from a manufacturing process. Unknown to the experimenters, items 1, 3, and 5 are good, while items 2 and 4 are defective. The procedure is to select one item from the five and subject it to test A. Then another item is selected from the remaining four and subjected to test B. Assuming that each of the $5 \times 4 = 20$ possible combinations of selections is equally likely to occur (with probability $\frac{1}{20}$), let $X =$ the item number tested by test A, and let $Y =$ the item number subjected to test B. Let E represent the event that the item chosen passes test A (i.e., that it is odd numbered) and let F represent the event that the item chosen passes test B. Thus $E = X$ odd and $F = Y$ odd. Then $P(E) = \frac{3}{5}$, $P(F) = \frac{3}{5}$, $P(E \cap F) = \frac{3}{10}$, and $P(E \cup F) = \frac{9}{10}$.

Indicator Random Variables. A random variable which can have only the values 0 and 1 is call an *indicator*. With a fixed sample space, there is a one-to-one correspondence between indicators and events: Given an event, the indicator is defined as having the value one at every sample point of the event and zero otherwise, and, given an indicator, an event is defined as consisting of all those sample points where the indicator takes the value one.

Indicators are useful in decomposing random variables into constituent parts corresponding to simple events. Suppose five blue tickets numbered from one to five are laid out on a table and covered by five red tickets with the same numbers, each red ticket covering exactly one blue ticket. Let X be the number of pairs with the same numbers, with sample space 1,2,3,4,5. The indicator variables X_n, $n=1,2,3,4,5$, are defined to be 1 if the nth blue ticket is covered by the nth red ticket, and zero otherwise. Then $X=X_1+X_2+X_3+X_4+X_5$. Some more important applications of indicators will occur subsequently.

1.6. Mean and Variance

The probabilities $P(X=x)=p_x$ which define a distribution contain all the information about the random variable X. Partial information can be obtained by calculating specially selected functions of the probabilities. The most useful functions of such kind are the *mean*,

$$m= \sum_{j=0}^{\infty} jp_j = \sum_{j=0}^{\infty} jP(X=j), \tag{7}$$

and the *variance*,

$$v= \sum_{j=0}^{\infty} (j-m)^2 p_j = \sum_{j=0}^{\infty} j^2 p_j - 2m \sum_{j=0}^{\infty} jp_j + m^2 \sum_{j=0}^{\infty} p_j$$

$$= \sum_{j=0}^{\infty} j^2 p_j - m^2. \tag{8}$$

(These summations have been written with $j=0,1,2,...$, which includes all the cases of concern in this chapter; naturally, if other values of the random variable occurred, the summations would have to be suitably modified.)

The various properties of the mean and variance are especially important in the study of statistics. Here it is sufficient to note that the mean is a kind of average value, where random variable values are weighted with the corresponding probabilities and summed, and that the variance is a measure of dispersion, since the weights are (squared) distances from the mean. A small variance indicates little variability and, in fact, zero variance indicates all probability concentrated at a single value $P(X=m)=1$. Such a distribution is called *causal* (since the cause of the phenomenon should be known) or *deterministic*.

In addition to the mean and variance, it is sometimes useful to consider *higher moments* m_r, $r=1,2,3,\ldots$, defined by the formula

$$m_r = \sum_{j=0}^{\infty} j^r p_j. \tag{9}$$

Hence $m=m_1$ is the first moment, and

$$v=m_2-m^2.$$

For probability distributions without a numerical sample space, the moments are not defined. Also, the definition of the moments depends on the connection (Section 1.4) between probabilities and sample points. It is therefore not quite right to speak of "the mean of the binomial distribution," because the binomial probabilities do not need to be assigned the conventional values $0,1,2,\ldots,n$, but could be arbitrarily associated with any set of $n+1$ values. Nevertheless, people do speak of the mean or variance of particular distributions, when it is clear from the context which values are assigned the probabilities of those distributions.

Sometimes it is more convenient to attribute the mean, variance, and higher moments to the random variable, using the notation $E(X), \text{var}(X)$ for mean and variance, respectively. This notation de-emphasizes the probabilities and is therefore useful when the particular distribution is fixed (or unspecified, simply p_x). Thus $m=E(X)$, $v=E(X^2)-[E(X)]^2$, $m_r=E(X^r)$, and so forth. The easiest way to remember and to become skillful in using this very handy notation is to regard $E(\ )$ as an abbreviation for $\sum p_x$. It has all the properties of a summation (and later an integral).

There are a few important cautions about $E(X)$. Students usually experience such a notation to mean a function; $f(x)$ represents a function of x. But $E(X)$ is not a function, but an operator on a function, and so the quantity on the other side of the equation will not involve X, but the parameters of the distribution. Also, $E(X)$ is called the *expected value* of X, and this can be extremely misleading, partly because $E(X)$ may not even be a value of X [$E(X)$ for the throw of a fair die is $3\frac{1}{2}$, but this is not a possible result], but also because even if it is a value of X, it is not necessarily the most probable value.

Example 1. Let $p_0 = \frac{3}{11}$, $p_1 = \frac{2}{11}$, $p_3 = \frac{1}{11}$, $p_4 = \frac{2}{11}$, and $p_5 = \frac{3}{11}$. Then $E(X) = 3$, the least probable value of X.

The ease of calculation with $E(X)$ is illustrated in evaluating the expression $\Sigma(x-c)^3 p_x$, where c is a constant:

$$E(X-c)^3 = E(X^3 - 3X^2c + 3Xc^2 - c^3)$$

$$= m_3 - 3cm_2 + 3c^2m - c^3.$$

In this calculation, no reference is made to p_x or to the values over which x is summed. Since E obeys the rules of summation or integration,

$$E(X+Y) = E(X) + E(Y)$$

and

$$E(cX) = cE(X).$$

There is one further important thing to remember about moments of distributions: They may not exist. In distributions which have an infinite number of terms, such as the Poisson distribution, the geometric distribution, the Fisher distribution, and many others which will be encountered later in the book, the calculation of $E(X)$ involves summing an infinite series which may converge or diverge. If the series diverges, the mean is said to be infinite, or to not exist. Similarly, when the mean does exist, some of the higher moments may not. Probability distributions with infinite moments are not exactly common, but neither are they artificial freaks of solely theoretical interest—they do occur in certain realistic situations. Here is how such a distribution is easily constructed.

Example 2 (A Distribution with an Infinite Mean). It is known that the series $\Sigma_{j=1}^{\infty}(1/j^n)$ converges for $n=2$, but diverges for $n=1$ (the well-known harmonic series). This means that if terms like $1/x^2$ are used for probabilities, the sum for the mean will involve terms like $1/x$, and will diverge.

Therefore define

$$p_1 = C,$$

$$p_2 = C/4,$$

$$p_3 = C/9,$$

$$p_4 = C/16,$$

$$\vdots$$

and, in general,

$$p_x = C/x^2, \qquad x = 1, 2, 3, \ldots,$$

where the constant C will be determined by normalization to unity:

$$1 = \sum p_j = C\left(1 + \tfrac{1}{4} + \tfrac{1}{9} + \tfrac{1}{16} + \cdots\right) = C\pi^2/6,$$

so that

$$p_x = \frac{6}{\pi^2 x^2}, \qquad x = 1, 2, 3, \ldots,$$

where of course π does not represent a parameter, but the Archimedean constant. By construction, $E(X) = \sum j p_j = \infty$.

By using the same series with $n = 3$, a distribution can be constructed with a finite mean but an infinite second moment, and with higher values of n, a distribution with the first $n - 2$ moments finite and infinite higher moments.

The idea of a probability distribution with an infinite mean value, although simple enough mathematically, has a history of confusion. One of the most famous "paradoxes" in probability, the *Petersburg Paradox* (Chapter 2, Problem 71) depends only on the failure to grasp the concept of such a probability distribution.

1.7. Calculation of the Mean and Variance

Given a sample space, a set of probabilities, and a rule connecting the two together, it is, in principle, not difficult to derive expressions for the moments in terms of the parameters, at least in many common distributions.

In this section, the necessary calculations are illustrated in several cases, calling attention to a few peculiarities and special techniques.

Consider first the Poisson distribution and its most frequent set of values, the non-negative integers:

$$P(X=x)=\frac{\lambda^x e^{-\lambda}}{x!}, \qquad x=0,1,2,\ldots,$$

The mean is defined to be

$$E(X)=\sum_{j=0}^{\infty} j\frac{\lambda^j e^{-\lambda}}{j!}.$$

In evaluating this sum, the first thing that springs to mind is to cancel j from numerator and denominator. There is a slight pitfall here, which has been known to trap students: The cancellation does not apply to the first term, which, in fact, vanishes. The first step is therefore

$$E(X)=\sum_{j=1}^{\infty} \frac{\lambda^j e^{-\lambda}}{(j-1)!}.$$

In order to reduce this to a standard exponential sum, it is necessary to align the exponent on λ with the factorial by removing λ to the front of the summation:

$$E(X)=\lambda e^{-\lambda}\sum_{j=1}^{\infty} \frac{\lambda^{j-1}}{(j-1)!}$$

$$=\lambda e^{-\lambda}\sum_{j=0}^{\infty} \frac{\lambda^j}{j!}$$

$$=\lambda.$$

This basically simple calculation is given in some detail because it illustrates features of technique which are quite often used. Another device is shown in the calculation of the second moment:

$$E(X^2)=\sum_{j=0}^{\infty} j^2\frac{\lambda^j e^{-\lambda}}{j!}.$$

Here, to provide the desired cancellation against the factorial, j^2 is written in the form $j(j-1)+j$; if the third moment is being computed, the substitution would be

$$j^3 = j(j-1)(j-2) + 3j^2 - 2j$$

and so forth, providing adequate lower-order terms to compensate for the factors required. In the quadratic case, two terms vanish in the first sum, leaving

$$E(X^2) = e^{-\lambda} \sum_{j=2}^{\infty} \frac{\lambda^j}{(j-2)!} + m$$

$$= \lambda^2 + \lambda,$$

so that the variance $=\lambda$.

Thus, when the Poisson distribution is defined over the non-negative integers, the mean and variance are both equal to the parameter λ.[†]

Finding the mean and variance of the geometric distribution over the non-negative integers involves another important technique: summing pre-multiplied geometric series. For example, if

$$P(X=x) = (1-\rho)\rho^x, \qquad x=0,1,2,\ldots, \ 0<\rho<1,$$

then

$$E(X) = \sum_{j=0}^{\infty} j(1-\rho)\rho^j.$$

To evaluate the sum, let

$$S = r + 2r^2 + 3r^3 + \cdots,$$

$$rS = r^2 + 2r^3 + \cdots,$$

$$(1-r)S = r + r^2 + r^3 + \cdots = r/(1-r),$$

$$S = r/(1-r)^2,$$

[†] This fact is sometimes taken by practical people in the reverse sense, assuming that data which exhibits equality of mean and variance must necessarily be from a Poisson distribution. This is a serious error, inasmuch as information about two moments is insufficient to determine all probabilities.

so that

$$E(X) = \rho/(1-\rho).$$

Another way to evaluate S is to notice that the coefficients make each term virtually a derivative:

$$\frac{S}{r} = \sum_1 jr^{j-1}$$

$$= \frac{d}{dr} \sum_1 r^j$$

$$= \frac{d}{dr}\left(\frac{r}{1-r}\right)$$

and so forth.

In finding the second moment, the series to be evaluated is

$$S = r + 4r^2 + 9r^3 + 16r^4 + \cdots.$$

Multiplying and subtracting twice according to the same scheme gives

$$rS = r^2 + 4r^3 + 9r^4 + \cdots,$$

$$(1-r)S = r + 3r^2 + 5r^3 + 7r^4 + \cdots,$$

$$r(1-r)S = r^2 + 3r^3 + 5r^4 + \cdots,$$

$$(1-r)^2 S = r + 2r^2 + 2r^3 + 2r^4 + \cdots$$

$$= r(1+r)/(1-r).$$

From this result, the values

$$E(X^2) = \frac{\rho(1+\rho)}{(1-\rho)^2},$$

$$\text{var}(X) = \frac{\rho}{(1-\rho)^2}$$

are easy to obtain.

With finite sums, such as the binomial over $0, 1, \ldots, n$, it is necessary to watch both ends of the summation for special cases:

$$E(X) = \sum_{j=0}^{n} j \binom{n}{j} \pi^j (1-\pi)^{n-j}.$$

The steps needed to reduce this to a standard binomial sum are similar to those used to reduce the Poisson mean to a standard exponential sum: cancellation of the factorial, adjustment of the index, and removal of the necessary constant:

$$E(X) = \sum_{j=1}^{n} \frac{n!}{(j-1)!(n-j)!} \pi^j (1-\pi)^{n-j}$$

$$= n \sum_{j=1}^{n} \binom{n-1}{j-1} \pi^j (1-\pi)^{n-j}$$

$$= n \sum_{j=0}^{n-1} \binom{n-1}{j} \pi^{j+1} (1-\pi)^{n-j-1}$$

$$= n\pi.$$

The student is strongly recommended to practice carrying out these calculations in various ways, for example, with the series written out to three or four terms in place of the summation sign, until the calculations are familiar and perhaps even automatic. Then it will be a good exercise to show that the variance of the binomial distribution is $n\pi(1-\pi)$.

1.8. The Distribution Function

The "exact" probabilities $P(X=x)$ completely describe a random variable; but so do many other functions. In fact, all that is needed for a function to work equally well is that it be obtainable from and transformable into the same exact probabilities. There are two categories of functions, each useful in its particular way, which serve this purpose: cumulative probabilities (this section) and generating functions (later integral transforms), beginning in Section 1.11.

Two types of cumulative probabilities will be introduced: the *distribution function*, defined as $P(X<x)$ and the *tail of the distribution*, defined by

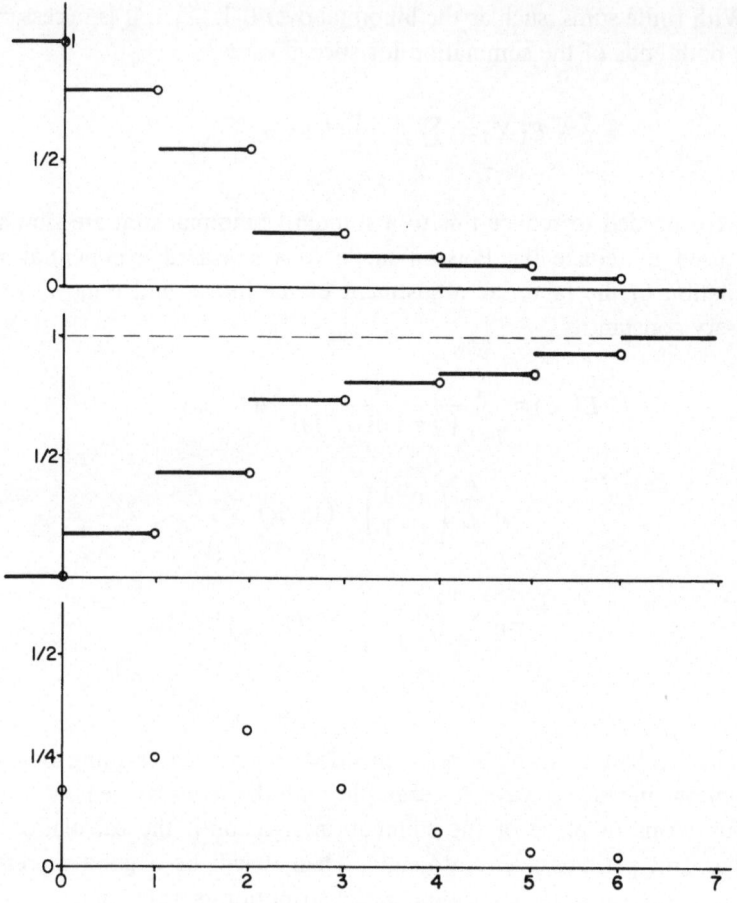

Figure 1. An illustrative probability distribution. Top, tail $Q(x)$; middle, distribution function $P(x)$; bottom, exact probabilities p_x.

$P(X \geq x)$.[†] The first important thing to notice is that whereas the exact probabilities are defined only for integer values of X (and it would be more complete always to add "zero elsewhere" in every equation), both of the cumulative distributions are, by virtue of the inequality sign, defined for all real values of x. This fact will be reflected in the notation by writing the argument in parentheses, rather than as a subscript, a common mathemati-

---

[†] These terms, which are quite standard, have two peculiarities: a lack of symmetry in name ("head" and "tail" might be better) and some awkwardness in using the word "distribution," both for the entire concept and for one particular function characterizing it.

cal indication,

$$P(x)=P(X<x),$$

$$Q(x)=P(X\geq x),$$

so that $P(x)+Q(x)=1$ for every value of x.

The cumulative functions are illustrated in comparison with the exact probabilities in Fig. 1; it is clear why $P(x)$ and $Q(x)$ are called "step functions." It can also be seen that the cumulative functions are discontinuous at points where $p_x \neq 0$ (shown in the diagram as the non-negative integers, for illustration). The functional values at these points of discontinuity are emphasized by large dots, showing that both cumulative functions are "left continuous," a property which will assume some importance in Chapter 4.[†]

At the points of discontinuity, the cumulative functions can be expressed in terms of the exact probabilities:

$$P(0)=0,$$

$$P(x)=p_0+p_1+p_2+\cdots+p_{x-1}, \qquad x=1,2,3,\ldots,$$

$$Q(x)=p_x+p_{x+1}+p_{x+2}+\cdots, \qquad x=0,1,2,\ldots .$$

Conversely, it is possible to obtain the exact probabilities from either cumulative function by the formulas

$$p_x=P(x+1)-P(x), \qquad x=0,1,2,\ldots,$$

(10)

$$p_x=Q(x)-Q(x+1), \qquad x=0,1,2,\ldots .$$

Note. These formulas are written under the assumption that the probabilities occur at the non-negative integers; some modification would be needed in other cases.

[†]Many authors define the cumulative functions so that they are both right continuous:· $P(x)=P(X\leq x)$, $Q(x)=P(X>x)$. The reason for a preference for left continuity appears in Section 4.9.

The left continuity properties are written as follows:

$$\lim_{y \to x+} P(y) = P(x+1),$$

$$\lim_{y \to x+} Q(y) = Q(x+1),$$

$$\lim_{y \to x-} P(y) = P(x),$$

$$\lim_{y \to x-} Q(y) = Q(x),$$

where x is integral. Also,

$$\lim_{x \to \infty} P(x) = \lim_{x \to -\infty} Q(x) = 1,$$

$$\lim_{x \to -\infty} P(x) = \lim_{x \to \infty} Q(x) = 0.$$

Expressing $P(x)$ and $Q(x)$ explicitly for particular distributions is sometimes easy, but often difficult. The student will have no trouble in showing that for the geometric distribution on the non-negative integers

$$P(x) = 1 - \rho^x,$$

$$Q(x) = \rho^x \tag{11}$$

at the integers, with modification for nonintegral values as follows:

$$P(x) = 0, \qquad x \le 0,$$
$$P(x) = 1 - \rho^k, \quad k < x \le k+1, k = 1, 2, 3, \ldots.$$

Similarly, the distribution function for the rectangular distribution

$$P_x = 1/n, \qquad x = 1, 2, \ldots, n, n = 1, 2, 3, \ldots,$$

can be written

$$P(x) = 0, \qquad x \le 1,$$

$$P(x) = j/n, \qquad j < x \le j+1, j = 1, 2, \ldots, n-1,$$

$$P(x) = 1, \qquad x > n.$$

On the other hand, the cumulative forms of the Poisson and binomial distributions must be reserved for Chapter 4, since they involve higher transcendental functions.

1.9. The Gamma Function and the Beta Function

In this section some mathematics is developed which will be necessary in treating certain probability distributions.

For $n>0$, the *gamma function* is defined by the definite integral

$$\Gamma(n)=\int_0^\infty e^{-t}t^{n-1}\,dt. \tag{12}$$

The integral is easy to integrate by parts, giving

$$\Gamma(n)=(n-1)\Gamma(n-1). \tag{13}$$

Therefore, when n is a positive integer, $\Gamma(n)=(n-1)!$, $\Gamma(1)=1$.

For nonintegral values of n, $\Gamma(n)$ is a transcendental function and cannot be expressed in terms of elementary functions, except for a few special cases, such as the half-integers. $\Gamma(\frac{1}{2})$ (and hence all other half-integers) can be calculated as follows. Let $t=x^2$, so Eq. (12) becomes

$$\Gamma(n)=2\int_0^\infty e^{-x^2}x^{2n-1}\,dx, \tag{14}$$

so that

$$\Gamma(\tfrac{1}{2})=2\int_0^\infty e^{-x^2}\,dx.$$

Usually this integral is evaluated by writing it first as a double integral,

$$\left[\Gamma(\tfrac{1}{2})\right]^2=4\int_0^\infty\int_0^\infty e^{-(x^2+y^2)}\,dx\,dy,$$

and then transforming to polar coordinates: $x=r\cos\alpha$, $y=r\sin\alpha$. Then some calculation gives the result

$$\Gamma(\tfrac{1}{2})=\pi^{1/2}.$$

Using the recursion relation (13), a general formula for all half-integers is easily calculated:

$$\Gamma(\tfrac{1}{2}n) = \frac{\pi^{1/2}(n-1)!}{2^{n-1}[\frac{1}{2}(n-1)]!}. \tag{15}$$

Occasionally the factorial notation is used for nonintegral values; then $n!$ is understood to denote $\Gamma(n+1)$. Thus $(-\tfrac{1}{2})! = \pi^{1/2}$, $\tfrac{1}{2}! = \tfrac{1}{2}\pi^{1/2}$, etc.

Another useful notation is *Pochhammer's symbol* for the product of x factors beginning with n, where the increase is by unit jumps:

$$(n)_x = n(n+1)(n+2)\cdots(n+x-1), \qquad x=1,2,3,\ldots.$$

In this formula, x must be a positive integer, but n can be any real number. Note that $(n)_x$ vanishes for $n=0, -1, -2, \ldots, 1-x$. It is easy to write $(n)_x$ as a quotient of gamma functions:

$$(n)_x = \frac{\Gamma(n+x)}{\Gamma(n)}. \tag{16}$$

The gamma function can also be used to extend the definition of the binomial coefficients $\binom{n}{x}$, since

$$\binom{n+x-1}{x} = \frac{\Gamma(n+x)}{\Gamma(n)x!} = \frac{(n)_x}{x!}, \qquad x=1,2,3,\ldots.$$

With this definition, it is possible to define $\binom{n}{x}$ for negative n:

$$\binom{-n}{x} = \frac{1}{x!}(-n-x+1)_x$$

$$= \frac{1}{x!}(-n-x+1)(-n-x+2)\cdots(-n)$$

$$= \frac{1}{x!}(-1)^x n(n+1)\cdots(n+x-1)$$

$$= (-1)^x \binom{n+x-1}{x}. \tag{17}$$

It is with this definition that the generalized binomial theorem

$$(A+B)^{-n}= \sum_{j=0}^{\infty} \binom{-n}{j} A^j B^{-n-j}$$

is valid. (For a proof, the student is referred to a textbook of algebra.)
The *beta function* is defined by the integral

$$B(p,q)= \int_0^1 x^{p-1}(1-x)^{q-1}\,dx.$$

This integral converges for $p>0$, $q>0$, and obviously $B(p,q)=B(q,p)$. Letting $x=\sin^2 \alpha$ gives the form

$$B(p,q)=2\int_0^{\pi/2}(\sin \alpha)^{2p-1}(\cos \alpha)^{2q-1}\,d\alpha.$$

There is an important relationship between the beta function and the gamma function:

$$\Gamma(p)\Gamma(q)=4\int_0^{\infty}\int_0^{\infty} x^{2p-1}y^{2q-1}e^{-(x^2+y^2)}\,dx\,dy.$$

Changing to polar coordinates gives

$$\Gamma(p)\Gamma(q)=4\int_0^{\pi/2}(\cos \alpha)^{2p-1}(\sin \alpha)^{2q-1}\,d\alpha \int_0^{\infty} r^{2(p+q)-1}e^{-r^2}\,dr$$

$$=\Gamma(p+q)B(p,q). \tag{18}$$

1.10. The Negative Binomial Distribution

This distribution has considerable importance both in probability and in statistical applications. It is based on the terms of a binomial series with negative exponents, and so contains an infinite number of values. These probabilities are usually assigned either to the non-negative integers (as in this section) or else to the positive integers beginning with a fixed number n. It contains the geometric distribution as a special case ($n=1$), but has little relationship with the binomial distribution except for one important application (Section 2.4).

Unlike most other elementary distributions, the negative binomial distribution appears in at least three rather different looking standard forms. The first is obtained simply by assigning the coefficients in the general binomial theorem (with $A = 1 - B$ for normalization to unity) to the non-negative integers:

$$P(X=x) = \left(\begin{matrix} -n \\ x \end{matrix} \right) A^x (1-A)^{-n-x}, \qquad x = 0, 1, 2, \ldots, \quad n = 1, 2, 3, \ldots .$$

For this assignment to produce a valid probability distribution, it is also necessary that all the terms be positive. This requires that the parameter A be negative, as will be more evident in the next standard form. Using Eq. (17),

$$P(X=x) = (-1)^x \left(\begin{matrix} n+x-1 \\ x \end{matrix} \right) A^x (1-A)^{-n-x}.$$

Finally, replacing the parameter A by the parameter ρ, where

$$(1-A)(1-\rho) = 1$$

gives the standard form

$$P(X=x) = \left(\begin{matrix} n+x-1 \\ x \end{matrix} \right) \rho^x (1-\rho)^n, \qquad x = 0, 1, 2, \ldots, \tag{19}$$

where n is a positive integer and $0 < \rho < 1$.

The moments of the negative binomial distribution can be found in much the same way as those of the binomial distribution. For example,

$$E(X) = \sum_{j=0}^{\infty} j \left(\begin{matrix} j+n-1 \\ j \end{matrix} \right) (1-\rho)^n \rho^j$$

$$= \sum_{j=1}^{\infty} \frac{(j+n)!}{(j-1)!(n-1)!} (1-\rho)^n \rho^j$$

$$= \sum_{j=0}^{\infty} \frac{(j+n)!}{j!(n-1)!} (1-\rho)^n \rho^j$$

$$= \frac{n\rho}{1-\rho} \sum_{j=0}^{\infty} \frac{(j+n)!}{j!n!} (1-\rho)^{n+1} \rho^j,$$

and this, being a binomial sum with parameter $n+1$, equals unity. Therefore

$$m = \frac{np}{1-p}.$$

1.11. The Probability Generating Function

One of the most spectacularly useful functions which characterizes a discrete distribution is the probability generating function. In addition to a number of unusually convenient properties which will be discussed later in this section, it has the initial merit of tidiness. A probability generating function combines the values of a random variable with the exact probabilities into a single function, so there is no need for a ragged list of values of x and range(s) of parameter(s) after each formula. (Here it is important to assume non-negative integral values.)

The probability generating function is a polynomial (or in the infinite case a power series) in which the coefficients are the probabilities and the exponents are the values of x. For example, with the distribution

$$P(X=0) = \tfrac{1}{2},$$

$$P(X=1) = \tfrac{1}{4},$$

$$P(X=2) = \tfrac{1}{4},$$

the probability generating function would be (with customary variable s)

$$\tfrac{1}{2} + \tfrac{1}{4}s + \tfrac{1}{4}s^2.$$

With the same probabilities, but different values of X, say

$$P(X=3) = \tfrac{1}{2},$$

$$P(X=4) = \tfrac{1}{4},$$

$$P(X=7) = \tfrac{1}{4},$$

the probability generating function would be

$$\tfrac{1}{2}s^3 + \tfrac{1}{4}s^4 + \tfrac{1}{4}s^7.$$

    With a random variable taking values on the non-negative integers, the probability generating function, denoted by $\phi(s)$, is defined by

$$\phi(s) = \sum_{j=0}^{\infty} P(X=j)s^j,$$

or, in other notation,

$$\phi(s) = p_0 + p_1 s + p_2 s^2 + p_3 s^3 + \cdots. \tag{20}$$

For virtually all the elementary distributions, the probability generating function (p.g.f.) can be expressed in simple form.

Rectangular Distribution   (with values $1, 2, \ldots, n$)

$$\phi(s) = s/n + s^2/n + s^3/n + \cdots + s^n/n$$

$$= \frac{s(1-s^n)}{n(1-s)}.$$

Poisson Distribution

$$\phi(s) = \sum_{j=0}^{\infty} \frac{e^{-\lambda}\lambda^j}{j!} s^j = e^{-\lambda + \lambda s}.$$

Binomial Distribution

$$\phi(s) = \sum_{j=0}^{n} \binom{n}{j} \pi^j (1-\pi)^{n-j} s^j$$

$$= (1-\pi)^n \sum_{j=0}^{n} \binom{n}{j} \left(\frac{s\pi}{1-\pi} \right)^j$$

$$= (1 - \pi + s\pi)^n.$$

Negative Binomial Distribution

$$\phi(s) = \frac{(1-\rho)^n}{(1-\rho s)^n}.$$

Geometric Distribution

$$\phi(s)=(1-\rho)/(1-\rho s).$$

Fisher Distribution

$$\phi(s)=\frac{\log(1-\beta s)}{\log(1-\beta)}.$$

In every case the proof is little more than a small variant on the sum which originally defined the probability. Unfortunately, there will be some distributions for which the sum is not quite so simple; but even if the p.g.f. is left in summation form, it can still be used and be useful.

One difficulty which students often have in understanding p.g.f.'s is knowing the meaning of s. This difficulty is not confined to students, for there is some discussion among mathematicians as to whether s should be regarded as a "label," a real variable, or something else.[†] For the purposes of this book, it is best to regard s as a real variable, defined at least over the interval $0 \leq s \leq 1$, and $\phi(s)$ as a function having rather interesting properties.

The first of these properties is that $\phi(1)=1$; this equation is the p.g.f. version of normalization to unity. In other branches of mathematics, generating functions are defined for an arbitrary sequence of constants $a_0+a_1 s+a_2 s^2 + \cdots$; the distinguishing feature of a p.g.f. is that since the coefficients are probabilities, setting $s=1$ must yield $\phi(1)=1$. What is the value of $\phi(0)$?

Since $\phi(s)$ contains all the information about the distribution, it must be possible to recover the exact probabilities. This, it is clear from Eq. (20), can be done by differentiation and setting $s=0$:

$$p_x = \phi^{(x)}(0)/x! \qquad (21)$$

Repeated differentiation with $s=1$ gives the moments of the distribution, with a little algebraic manipulation:

$$\phi'(s)=p_1+2sp_2+3s^2p_3+4s^3p_4 + \cdots,$$

so that

$$m=E(X)=\phi'(1),$$

[†]A useful interpretation of s appears in Section 2.4 and subsequently.

a third expression for the mean. Furthermore,

$$\phi''(s)=2p_2+3.2sp_3+4.3s^2p_4+\cdots,$$

$$\phi''(1)=2p_2+3.2p_3+4.3p_4+\cdots,$$

which very much resembles

$$E(X^2)=p_1+2.2p_2+3.3p_3+4.4p_4+\cdots.$$

Subtracting these equations gives

$$E(X^2)-\phi''(1)=E(X).$$

Thus the variance can be expressed in terms of the p.g.f.:

$$v=m_2-m^2$$

$$=\phi''(1)+\phi'(1)-\left[\phi'(1)\right]^2.$$

In a similar fashion, the third and higher moments can be written in terms of the p.g.f. at the point $s=1$.

Change of Variable. A function of a random variable is itself a random variable. Specifically, two functions of importance in this book are adding a constant $(X+k)$ and multiplying by a constant (cX). Neither of these operations affects the probabilities, but both affect the values of the variables. Consider the example given at the beginning of this section, with $k=3$ or with $c=5$. In the first instance, the distribution (writing $Y=X+3$, p.g.f. ψ for Y) is

$$P(Y=3)=\tfrac{1}{2},$$

$$P(Y=4)=\tfrac{1}{4},$$

$$P(Y=5)=\tfrac{1}{4},$$

and

$$\psi(s)=s^3\phi(s).$$

In the second case, writing $Z=5X$, with p.g.f. $\chi(s)$,

$$P(Z=0)=\tfrac{1}{2},$$

$$P(Z=5)=\tfrac{1}{4},$$

$$P(Z=10)=\tfrac{1}{4},$$

and

$$\chi(s)=\phi(s^5).$$

The student can verify the general principle suggested by these examples: If random variables X, Y, and Z have p.g.f. $\phi_X(s)$, $\phi_Y(s)$, and $\phi_Z(s)$, then

$$Y=X+k \text{ is equivalent to } \phi_Y(s)=s^k\phi_X(s) \qquad (22)$$

and

$$Z=cX \text{ is equivalent to } \phi_Z(s)=\phi_X(s^c); \qquad (23)$$

that is, translation of a random variable corresponds to multiplication by s of the p.g.f., and multiplication of a random variable by a constant corresponds to raising the s variable to a power.

Generating Functions of Cumulative Probabilities. As an example of generating functions which are not p.g.f.'s, consider

$$\Phi(s)= \sum_{j=1}^{\infty} P(j)s^j,$$

$$\Psi(s)= \sum_{j=0}^{\infty} Q(j)s^j,$$

where, as in earlier notation, $P(x)=P(X<x)$, $Q(x)=P(X\geq x)$, $\phi(x)=\sum P(X=j)s^j$.

The following relationships between the generating functions of the cumulative probabilities and the p.g.f.'s can be obtained without much difficulty:

$$\Phi(s) = \frac{s\phi(s)}{1-s},$$

(24)

$$\Psi(s) = \frac{1-s\phi(s)}{1-s}.$$

Note: Different generating functions can be denoted either by using different letters (as above) or by indicating the variable as a subscript [as in Eqs. (22) and (23)].

Since Φ and Ψ equally characterize the probability distribution, the mean can be expressed in terms of these generating functions. Differentiating Eq. (24) and setting $s=1$ gives the formula

$$E(X) = \Psi(1) - 1.$$

Similarly,

$$v = 2\Psi'(1) + \Psi(1) - [\Psi(1)]^2.$$

1.12. The Catalan Distribution

Generating functions are useful in many ways. In this section an example is given in which the generating function is easier to find than the coefficients which define it, so that the coefficients are calculated from the generating function, rather than the other way around.

The problem is not probabilistic, at least not in the beginning, although a probabilistic meaning subsequently appears. It concerns the order in which an algebraic operation such as multiplication of several factors can be performed. The expression abc means the product of these three quantities, but it is possible to multiply first ab and then the result by c, or else bc and then the result by a. To indicate order, parentheses are used: $(ab)c$ for the first order, and $a(bc)$ for the second. [It is assumed that the product is only between adjoining symbols, excluding $(ac)b$, for instance.]

With three symbols, the number of ways of multiplying is clearly two;

this is indicated by writing $w_3 = 2$. It is obvious that $w_2 = 1$ and $w_1 = 1$. To find w_4, enumeration is simplest:

$$a(b(cd)), \quad (a(bc))d, \quad a((bc)d), \quad (ab)(cd), \quad ((ab)c)d;$$

therefore $w_4 = 5$. Similar experimentation will give the values $w_5 = 14$, $w_6 = 42$.

Is there a general formula for $w_k =$ the number of ways to insert parentheses in a product of k factors? The first step in finding such a formula is to establish a recurrence relation for calculating successive values. Suppose that the last multiplication in the sequence is between j factors on the left and $k - j$ factors on the right, where $j = 1, 2, 3, \ldots, k - 1$. The number of ways the earlier multiplications on the left could have been performed is by definition w_j, and on the right, w_{k-j}.

Hence the number of ways w_k can be written is

$$w_k = w_1 w_{k-1} + w_2 w_{k-2} + \cdots + w_{k-2} w_2 + w_{k-1} w_1, \tag{25}$$

and w_k can be calculated step by step from this formula. In fact, the student will be able, using the recursion formula, to find w_k up to a dozen or so factors without much trouble, especially the student with a small calculator.

This still does not give a general formula for w_k as a function of k, and it is a real challenge to the ingenuity to try to convert the recursion relation into such a formula. One of the simplest methods is by use of the generating function

$$\phi(s) = \sum_{j=1}^{\infty} w_j s^j.$$

Here is how to do it: Multiply the generating function by itself:

$$[\phi(s)]^2 = w_1^2 s^2 + (w_2 w_1 + w_1 w_2) s^3 + (w_1 w_3 + w_2 w_2 + w_3 w_1) s^4 + \cdots.$$

It is immediately obvious that the coefficients in parentheses can be simplified by using the recurrence formula:

$$[\phi(s)]^2 = w_2 s^2 + w_3 s^3 + w_4 s^4 + \cdots$$

$$= \phi(s) - s. \tag{26}$$

This is a simple quadratic equation in the generating function, namely $\phi^2 - \phi + s = 0$, with solutions

$$\phi(s) = \tfrac{1}{2}\left[1 \pm (1-4s)^{1/2}\right].$$

The negative sign is chosen so that the condition $\phi(0) = 0$ is satisfied, and s must be less than $\tfrac{1}{4}$ for real roots.

The next problem is to find the w_k from $\phi(s)$. Since the w_k are the coefficients in the power-series expansion of $\phi(s)$, it is only necessary to express $\phi(s)$ in powers of s. Using the binomial theorem,

$$(1-4s)^{1/2} = \sum_{j=0}^{\infty} \binom{\tfrac{1}{2}}{j}(-4s)^{j}.$$

Therefore

$$\phi(s) = -\frac{1}{2}\sum_{j=1}^{\infty} \binom{\tfrac{1}{2}}{j}(-4s)^{j},$$

and w_k is found by lifting out the coefficient of s^k:

$$w_k = (-1)^{k-1} 2^{2k-1}\binom{\tfrac{1}{2}}{k}.$$

The most troublesome part of the calculation is to simplify this expression:

$$w_k = (-1)^{k-1} 2^{2k-1}\frac{1}{k!}\left[(\tfrac{1}{2})(-\tfrac{1}{2})(-\tfrac{3}{2})(-\tfrac{5}{2})\cdots(\tfrac{3}{2}-k)\right]$$

$$= 2^{2k-1}\frac{1}{k!}\left[(\tfrac{1}{2})(\tfrac{1}{2})(\tfrac{3}{2})\cdots(k-\tfrac{3}{2})\right]$$

$$= 2^{k-1}\frac{1}{k!}\left[1\cdot 3\cdot 5\cdots(2k-3)\right]$$

$$= \frac{1}{k}\binom{2k-2}{k-1}$$

or, in the alternative form,

$$w_k = \frac{1}{k-1}\binom{2k-2}{k}.$$

Now the values obtained already by enumeration or recurrence can be confirmed by direct substitution.

Remember that this sequence does not form a probability distribution [$\phi(1) \neq 1$]. Nevertheless, it can be used to construct a probability distribution by the familiar method of normalization to unity. Dividing by $\phi(s)$,

$$\frac{w_x s^x}{\phi(s)}, \qquad x=1,2,3,\ldots,$$

is a valid probability distribution with parameter ϕ (or s, since they are related by the quadratic equation). In Chapter 6 this distribution occurs as an important one in the theory of queues, with a parameter ρ satisfying

$$\phi = \frac{\rho}{1+\rho}.$$

Using Eq. (26), it is easy to see that

$$s = \frac{\rho}{(1+\rho)^2}$$

and that the probabilities can be written in terms of the parameter ρ as follows:

$$P(X=x) = \frac{1}{x}\binom{2x-2}{x-1}\left(\frac{\rho}{1+\rho}\right)^{x-1}\left(\frac{1}{1+\rho}\right)^x, \qquad x=1,2,3,\ldots. \quad (27)$$

The distribution will be called the Catalan[†] distribution, since the numbers w_k are known to mathematicians as the Catalan numbers.

1.13. More about the p.g.f.; The Equation $s = \phi(s)$

In Section 1.11, two properties of the p.g.f., $\phi(1)=1$ (normalization to unity) and $\phi'(1)=E(X)$, were discussed. For a deeper understanding, it is helpful to consider the equation $t=\phi(s)$ in the s–t plane, assuming, as usual, non-negative integral values for X. Since $\phi(0)=P(X=0)$, the two ends of the curve are fixed at $(0, p_0)$ and $(1, 1)$. Also, the fact that all coefficients p_x are positive means that the curve is continuous, monotonically increasing, with a monotonically increasing slope.

[†]Eugene Charles Catalan, Belgian mathematician, 1814–1894.

The only exceptions are straight lines, and it may be well to deal with these cases first. The right end being anchored at $(1,1)$, there are two cases:

(a) $t=\phi(s)$, a horizontal line, therefore $p_0=1$, and the distribution is causal at the origin.

(b) $\phi(s)=p_0+p_1 s$, a diagonal line connecting $(0, p_0)$ and $(1,1)$. Then $p_1=1-p_0$, and the distribution is a simple version of the binomial, with $n=1$. This is called the Bernoulli[†] distribution and is modeled by a single throw of a (not necessarily fair) coin.

As a special case of (b), there is case (c), $p_1=1$, the causal distribution at the value $x=1$.

Except for these special cases, $t=\phi(s)$ is a curve which is at least quadratic. Furthermore, the slope at $(0, p_0)$ is p_1 (non-negative) and the slope at $(1,1)$ is m, positive except for the degenerate case (a).

In one of the applications of Markov chain theory in Chapter 3, it is important to know whether or not the curve $t=\phi(s)$ crosses the line $t=s$ between $(0, p_0)$ and $(1,1)$, i.e., to find whether the equation $s=\phi(s)$ has a root in the interval $(0,1)$. [It always has the root $s=1$ (normalization to unity once again).] Because of the increasing slope of the curve, there can be at most one such root [except for the degenerate case (c)].

If there is a root in this interval, the slope of the curve must be greater than unity at the crossing point and must therefore be greater than unity at $(1,1)$. Thus $m>1$. On the other hand, if there is no such root, $t=\phi(s)$ will always be above $t=s$ [except for case (c)] and so the slope at its largest value $\phi'(1)$ will be less than or equal to one. This shows that, except for the trivial cases, the existence of a root of the equation $\phi(s)=s$ in the interval $0<s<1$ is a necessary and sufficient condition for $m>1$.

As an example, consider the Poisson distribution with parameter λ. The equation $\phi(s)=s$ can be written

$$(e^\lambda)^s=se^\lambda$$

or

$$\lambda=\frac{\log s}{s-1}.$$

Looking at the function $\log s$, it is easy to see that $\lambda=1$ only where $s=1$, and otherwise $s-1>\log s$. Thus, if s is greater than unity, $s-1$ is positive,

---

[†]Jacob Bernoulli, Swiss mathematician, 1654–1705.

and $\log s/(s-1)$ will be less than one. Reversing the inequality on s reverses it on λ.

1.14. Problems

1. For discussion of sample spaces:
    - (i) The amount of money carried by a randomly chosen person
    - (ii) The number of teeth of a randomly chosen person
    - (iii) The number of students in a class
    - (iv) The number of axles on a vehicle
    - (v) The number of working light bulbs in a house
    - (vi) The number of books published by an author
    - (vii) The number of ways of going between two places
    - (viii) The number of countries bordering on a country
    - (ix) The number of area codes (or zip codes) in a state
    - (x) The number of windows in a building

2. For a Poisson distribution with parameter λ, show that the largest probability corresponds to x, the largest integer less than or equal to λ.

3. What is the largest probability for the binomial distribution?

4. Make the first three terms of an arithmetic progression into a probability distribution by normalizing to unity. How many parameters will there be in the distribution? What range of values can the parameters have?

5. In Section 1.5, Example 2, find the probability that X is an odd number.

6. In Section 1.5, Example 3, find the probability that there will be more than three claims in one day.

7. In Section 1.5, Example 4, find the probability that X is odd, square, and less than ten.

8. In a Poisson distribution on the non-negative integers, $P(X=1)=P(X=2)$. Find $P(X=4)$. $Ans.\ \frac{2}{3}e^{-2}$.

9. If X is binomially distributed with parameters n and π, show that (i) $n-X$ is binomially distributed with parameters n and $1-\pi$, (ii) the distribution is symmetrical about the value c if and only if $\pi=\frac{1}{2}$, $c=\frac{1}{2}n$.

10. Find the mean, variance, and probability generating function for the geometric probabilities defined for the following values: (i) $1,2,3,\ldots$, (ii) $2,3,4,\ldots$, (iii) $k, k+1, k+2, \ldots$, (iv) $1,3,5,\ldots$.

11. Let E and F be events, and let E' be the event consisting of all sample points not in E. Let I_E, I_F, and $I_{E'}$ be the respective indicators. Show that (i) $I_{E'}=1-I_E$, (ii) $I_{E\cap F}=I_E I_F$, (iii) $I_{E\cup F}=1-(1-I_E)(1-I_F)$.

12. Using the indicator relationship of Problem 11, show that for three events E, F, and G, $E \cup (F \cap G) = (E \cup F) \cap (E \cup G)$.

13. Generalize the red and blue ticket problem of Section 1.5 to n tickets of each color. Using indicators, show that $E(X) = 1$, independent of the number of ticket pairs.

14. Let X be a random variable with a geometric distribution over the non-negative integers, with parameter $\rho = \frac{1}{2}$. Compute the probabilities (i) $X > 4$, (ii) $3 \leq X < 6$, (iii) $4 < X \leq 6$ or $X > 8$, (iv) $2 \leq X \leq 4$ or $8 < X < 12$.

$$\text{Ans. (i) } \tfrac{1}{32}, \text{ (ii) } \tfrac{7}{64}, \text{ (iii) } \tfrac{13}{512}$$

15. Let X be Poisson distributed over the non-negative integers with parameter λ. Find the expected value of $(1 + X)^{-1}$. $\qquad\qquad$ Ans. $(1 - e^{-\lambda})/\lambda$

16. Prove that the variance of aX is equal to $a^2 v$, where v is the variance of X.

17. Let X be a binomial variable with parameters π and n. Find the expected value of $(1 + X)^{-1}$.

18. A random variable has distribution

$$P(X = x) = \beta(1 - \beta)^x + \beta(1 - \beta)^{x-1}, \qquad x = 1, 3, 5, \ldots, \; 0 < \beta < 1.$$

Show that this normalizes to unity and find $E(X)$.

19. For the Poisson distribution over the non-negative integers with parameter λ, find $E[(X - \lambda)^3]$.

20. Which distribution defined over $1, 2, \ldots, n$ has the largest variance?

21. Let X be a random variable defined over $N, N+1, N+2, \ldots$, where N is a positive integer. Suppose

$$P(X = x) = (1 - \rho)\rho^{x-N}, \qquad x = N+1, N+2, N+3, \ldots.$$

Find $P(X = N)$ and show that

$$E(X) = \frac{N - N\rho + \rho^{N+1}}{1 - \rho}.$$

22. Find the cumulative probabilities $P(x)$ and $Q(x)$ as defined in Section 1.8 for the distributions of Problem 10.

23. For a probability distribution over $N, N+1, N+2, \ldots$, where N is a positive integer, express $P(X = x)$ in terms of $P(x)$, in terms of $Q(x)$. Express $E(x)$ in terms of $Q(x)$. Are the formulas independent of N?

$$\text{Ans. } E(x) = N + \sum_{j=1}^{\infty} Q(N + j)$$

24. Let $p_x(\lambda)$ represent the Poisson probabilities for $x=0,1,2,\ldots$ and let $P(x,\lambda)$ and $Q(x,\lambda)$ be the cumulative probabilities (with parameter indicated). Prove the following formulas:

(i) $\displaystyle\int_0^\infty P(x-1,\lambda t)\,dt=x/\lambda,$

(ii) $\displaystyle\sum_{j=0}^x p_j(\lambda)p_{x-j}(\mu)=p_x(\lambda+\mu),$

(iii) $\displaystyle\int_0^u p_x(\lambda t)\,dt=Q(x,\lambda u),$

(iv) $\displaystyle\int_0^u Q(x-1,\lambda t)\,dt=(1/\lambda)\sum_{j=x}^\infty Q_j(\lambda u),$

(v) $\displaystyle\frac{d}{d\lambda}Q(x,\lambda)=p_x(\lambda).$

25. Let $p_x=P(X=x)$ be a discrete probability distribution over the non-negative integers, and let $f(x)=\sum_{j=1}^x jp_j$ and $g(x)=\sum_{j=0}^x p_j$. Show that

$$f(x)=f(\infty)g(x-1)$$

if (and only if) the distribution is Poisson.

26. Show that

$$\int_0^\infty \frac{y^{b-1}}{1+y}\,dy=\Gamma(1-b)\Gamma(b).$$

27. Show that

$$\Gamma(n)=2\int_0^\infty y^{2n-1}e^{-y^2}\,dy.$$

28. Show that

$$\Gamma(n)=\int_0^1 [\log(1/t)]^{n-1}\,dt.$$

29. Find the variance of the negative binomial distribution directly, using the p.g.f.

30. Referring to Section 1.10, draw the graph of $(1-A)(1-\rho)$ in the $A-\rho$ plane and verify that the statements about the values of these parameters are correct.

31. Let $P(X=x)=\frac{1}{6}x$, $x=1,2,3$. Find $P(X^2=x)$.   Ans. $x^{1/2}/6$

32. Let X be geometrically distributed over the non-negative integers with parameter ρ. Let $Y=\min(X, M)$, where M is an integer. Compute the probability distribution of Y. Work the same problem where X takes values in the positive integers.

33. Let X be geometrically distributed with parameter ρ. Compute the following distributions: (i) X^2 if X takes values in the non-negative integers; (ii) X^2 if X takes values in the positive odd integers; (iii) $X+4$ if X takes values in the non-negative integers; and (iv) $X+4$ if X takes values in the positive odd integers.

34. Let X be geometrically distributed over the non-negative integers with parameter ρ, and let K be an integer. Find $E(\min X, K)$ and $E(\max X, K)$. Hint: Use $E(X)=\Sigma Q(j)$, Problem 37.

35. With the assumptions of Problem 34, find the distribution of $Y=\min(X, K)$.

36. Let X be a random variable satisfying

$$P(X=x)=\frac{2x}{N(1+N)}, \qquad x=1,2,\ldots, N.$$

Show that this is a valid probability distribution, and find the mean, variance, and p.g.f.

37. Referring to Section 1.11, show without use of the p.g.f. that $E(X)=\Sigma Q(j)$ if $x=0,1,2,\ldots$ (compare Problem 23).

38. A random variable has p.g.f. $(3+s)/(6-2s)$. Find the mean, variance, and $P(X=x)$. \qquad *Ans.* $P(X=x)=3^{-x}, x=1,2,3,\ldots, P(x=0)=\frac{1}{2}$.

39. In Problem 4, find the mean, variance, and p.g.f. for the following values of X: (i) $0,1,2$; (ii) $2,3,4$; (iii) $1,5,6$.

40. Show that the generating function of (i) a^x is $(1-as)^{-1}$, (ii) x is $s/(1-s)^2$, (iii) $x(x-1)$ is $2s^2/(1-s)^3$, (iv) x^2 is $s(s+1)/(1-s)^3$.

41. The *moment generating function* is defined to be $E(e^{sX})$. Show that it actually generates (i.e. has coefficients) the moments divided by factorials. Find a formula connecting the moment generating function with the p.g.f.

42. Find the mean, variance, and p.g.f. for the following:
    (i) The Poisson probabilities assigned to the positive odd integers, i.e.,

    $$P(X=x)=\frac{e^{-\lambda}\lambda^{(x-1)/2}}{[(x-1)/2]!}, \qquad x=1,3,5,\ldots.$$

    (ii) The odd Poisson probabilities normalized to unity over the same odd values, i.e.,

    $$P(X=x)=C\frac{e^{-\lambda}\lambda^x}{x!}, \qquad x=1,3,5,\ldots.$$

43. The concept of the p.g.f. of a random variable can be generalized to that of a generating function of an event, as $\Phi(s)$ and $\Psi(s)$ of Section 1.11, where the events in question were $X<x$ and $X\geq x$. Find the generating functions for the following events, where $\phi(s)$ is the p.g.f. of X: (i) $X>x+1$, (ii) $X\leq x$, (iii) $X>x$, (iv) $X=2x$.

44. Given that $\frac{1}{6}(2s+1)(1+s)$ is the p.g.f. for a random variable X, find (i) $P(X=x)$, (ii) the p.g.f. for $X+1$, (iii) $E(X)$, (iv) the variance of X.

45. Find $P(x)$ and $Q(x)$ for Problems 31, 36, and 38.

46. Given a random variable with $\phi(s)=\frac{1}{10}(s+1)(2s+3)$, find the mean, variance, and $P(x)$.

47. Let $\phi(s)$ be a p.g.f. (i) show that $1/[2-\phi(s)]$ is also a p.g.f. (ii) If the values of the random variable corresponding to $\phi(s)$ are the non-negative integers, what are the values of the random variable in part (i)? (iii) If $\phi(s)$ corresponds to a random variable X and part (i) to a random variable Y, express $P(Y=0)$ in terms of the probabilities of X. (iv) Show that $E(X)=E(Y)$.

48. Consider the probability distribution $P(X=x)$, with p.g.f. $\phi(s)$. Define the *exponential probability generating function* $\psi(s)$ by

$$\psi(s)=\sum_{j=0}^{\infty} P(X=j)\frac{s^j}{j!}.$$

Show that

$$\phi(s)=\int_{0}^{\infty}e^{-s}\psi(ts)\,dt.$$

49. Random variables X and Y are connected through their p.g.f., and $\phi(s)$ and $\psi(s)$, by the relation $s^2\phi(s)=\psi(s)$. Find the relationships between $E(X)$ and $E(Y)$ and between the variances.

50. Let $p_x(\lambda)$ denote the Poisson probabilities over the non-negative integers. Define a probability distribution $q_x(\lambda)$ [random variable X] by

$$q_x(\lambda)=p_{x-1}(\lambda), \qquad x=1,3,5,\ldots,$$

$$q_x(\lambda)=p_{x+1}(\lambda), \qquad x=0,2,4,\ldots.$$

Find $E(X)$.

51. Let $P(X=x)=(1-\rho)\rho^x$, $x=0,1,2,\ldots$ and let a distribution q_x be defined by

$$q_x=P(X=2x)+P(X=2x+1), \qquad x=0,1,2,\ldots.$$

Find the mean, variance, and p.g.f. for q_x.

52. Express $E(X^3)$ in terms of the p.g.f.   *Ans.* $\phi'''(1)+3\phi''(1)+\phi'(1)$.

53. Find $P(1\leq X\leq 2)$, where (i) X is Poisson with $\lambda=1$, (ii) X is binomial with $n=10$ and $\pi=0.1$, (iii) X is negative binomial with $\pi=0.1$ and $n=10$, (iv) X is geometric with $\rho=\frac{1}{2}$, (v) X is Catalan with $\rho=0.1$. In each case, the "natural" values are used, e.g., the non-negative integers for the Poisson distribution.

54. Show that the mean of the Catalan distribution is $(1+\rho)^{-1}$ and that its variance is $\rho(1+\rho)(1-\rho)^{-3}$.

55. Find the roots of the equation $s=\phi(s)$ for the geometric distribution with p.g.f. $(1-\rho)/(1-s\rho)$.

56. Show by Section 1.13 that the expected value of an indicator is less than unity.

57. Every indicator has a Bernoulli distribution; is the converse true?

58. A fair die is thrown and X is the sum of the number showing on top and on the face nearest you. Find the distribution of X.
    Note: Opposite faces of a die are numbered $1:6$, $2:5$, and $3:4$.
                     Ans. $P(X=x)=\frac{1}{6}$, $x=5,6,8,9$; $P(X=x)=\frac{1}{12}$, $x=3,4,10,11$.

59. In Section 1.7 (geometric distribution) use the "derivative" method, as shown for finding $E(X)$ to sum the series needed to find $E(X^2)$.

60. Nine points are arranged in a square, and three of them are chosen at random. Two points are "neighbors" if adjacent in a row, column, or diagonal. Find the distribution of $X=$ the number of neighbors.

61. A coin is weighted so that the probability of heads is $\frac{1}{4}$; it is tossed three times. Find the distribution of $X=$ the length of the longest sequence of tails observed. $X=0$ if no tails.   *Ans.* $\left(\frac{1}{64},\frac{18}{64},\frac{18}{64},\frac{27}{64}\right)$

62. A boy comes from a family with three children; what is the probability that his two siblings are of the same sex? Assume equal probability that a child is male or female.   *Ans.* $\frac{4}{7}$

63. Given a random variable X, with

$$P(X=2)=a(1-b), \qquad P(X=3)=ab, \qquad P(X=5)=1-a.$$

(i) Sketch the region in the a–b plane for which this is a valid probability distribution, (ii) find $P(2X-3<5)$, (iii) find $E(X)$, (iv) sketch the probability generating function.

64. A die is rolled three times; what is the probability that each result is larger than the preceding one?   *Ans.* $\frac{5}{54}$

65. Show that the recurrence equation (25) for Catalan numbers is satisfied if w_k is interpreted as the number of ways a convex polygon of $k+1$ sides can be divided into triangles by $k-2$ nonintersecting diagonals.

66. A green die and a red die are thrown (together) n times. Show that the number of times the value on the green die is greater than the value on the red die is a binomial random variable. Find the parameters.

2

Conditional Probability

2.1. Introduction. An Example

This chapter deals with probability distributions and their properties—very much as in Chapter 1—but with the additional feature that the total sample space is constrained by a "given" event. Frequently the event will consist of particular values imposed on a random variable. To prepare for the formal development of conditional probability, the following example will be instructive.

Suppose there are three fish swimming in a bowl, and a person tries to catch them by making repeated passes with a net. Assume further that the probability of catching X fish on the first attempt is binomial with parameters 3 and π, where π characterizes the efficiency of the netting process. That is, with $X=$ number of fish caught,

$$P(X=x)=\binom{3}{x}\pi^x(1-\pi)^{3-x}, \qquad x=0,1,2,3.$$

Next, suppose a second attempt is made to catch the fish, and the number caught the second time is the random variable Y. Again, suppose that the binomial distribution applies to Y, with one parameter equal to the number of fish remaining after the first trial, $3-X$.

Some reflection on the random variables X and Y should show that Y has also the possible values $0,1,2,3$—and yet there are certain combinations of values which are impossible, for example, $X=3$, $Y=3$. In other words, not all of the 16 possibilities have positive probability. In fact,

$$\text{if } X=0, \quad \text{then } Y=0,1,2,3,$$

$$\text{if } X=1, \quad \text{then } Y=0,1,2,$$

$$\text{if } X=2, \quad \text{then } Y=0,1,$$

$$\text{if } X=3, \quad \text{then } Y=0.$$

There are really 10 nonzero probabilities, and these are easy to write down, using the binomial assumption for definiteness.

If $X=0$ [the probability of which is $(1-\pi)^3$], then

$$\left.\begin{array}{l} P(Y=0)=(1-\pi)^3 \\ P(Y=1)=3\pi(1-\pi)^2 \\ P(Y=2)=3\pi^2(1-\pi) \\ P(Y=3)=\pi^3 \end{array}\right\} \quad \text{(binomial with parameters 3 and } \pi\text{)}.$$

If $X=1$ [the probability of which is $3\pi(1-\pi)^2$], then

$$\left.\begin{array}{l} P(Y=0)=(1-\pi)^2 \\ P(Y=1)=2\pi(1-\pi) \\ P(Y=2)=\pi^2 \end{array}\right\} \quad \text{(binomial with parameters 2 and } \pi\text{)}.$$

If $X=2$ [the probability of which is $3\pi^2(1-\pi)$], then

$$\left.\begin{array}{l} P(Y=0)=1-\pi \\ P(Y=1)=\pi \end{array}\right\} \quad \text{(binomial with parameters 1 and } \pi\text{)}.$$

If $X=3$ [the probability of which is π^3], then

$$P(Y=0)=1 \quad \text{(binomial with parameters 0 and } \pi\text{, i.e., causal)}.$$

These 10 probabilities are said to be *conditional on values of X*, and a vertical bar is used to represent the relationship:

$$P(Y=0|X=0)=(1-\pi)^3,$$

$$P(Y=1|X=0)=3\pi(1-\pi)^2,$$

$$P(Y=2|X=0)=3\pi^2(1-\pi),$$

$$P(Y=3|X=0)=\pi^3,$$

$$P(Y=0|X=1)=(1-\pi)^2,$$

$$P(Y=1|X=1)=2\pi(1-\pi),$$

$$P(Y=2|X=1)=\pi^2,$$

$$P(Y=0|X=2)=1-\pi,$$

$$P(Y=1|X=2)=\pi,$$

$$P(Y=0|X=3)=1.$$

[It would also be possible to write down the various possibilities which have probability zero, for example, $P(Y=3|X=3)=0$, etc.]

If the expression $P(Y=y|X=x)$ is regarded as a function of the two arguments x and y, it is clear that the normalization to unity applies to the first argument only: Summing on x is a meaningless operation.

These 10 probabilities form four separate probability distributions, one conditional on each of the four possible values of X, but they *do not represent the individual probabilities of the* 10 *basic sample points* which have positive probability. For example, although the value $Y=0$ conditional on $X=3$ has probability unity, it does not mean that the joint occurrence of $X=3$ and $Y=0$ is mandatory. The probabilities of the basic sample points are denoted by a comma,

$$P(X=x, Y=y),$$

in contrast with

$$P(X=x|Y=y).$$

There is a great difference between saying that two fish are caught the first

Table 2.1. Data for the fish-catching experiment

$Y=3$	0	0	0
$Y=2$		0	0
$Y=1$			0
$Y=0$			
$X=0$	$X=1$	$X=2$	$X=3$

time and none the second time, and saying that two fish are caught the first time, given that none are caught the second time.

How shall the probabilities of the sample points be calculated? First, consider the various possibilities arranged in a rectangular tableau: in the language of probability, a "bivariate distribution" or, in the language of statistics, a "contingency table." Just as the sample points and their probabilities are listed in the single-variable case, so, in the bivariate case, they are given in a two-way table. For the fish-catching experiment, the table would have 16 cells, six of them occupied by zeros, as in Table 2.1.

In thinking about how to fill in the remaining probability values in Table 2.1, note first that the column totals, i.e., $P(X=x)$, $x=0,1,2,3$, are given by the original assumption as binomial values. This says only that the probability of catching x fish the first time is the sum of the probabilities of the constituent sample points, $P(X=x, Y=y)$ summed on y.

Therefore the probability in the lower right cell must be $P(X=3)=\pi^3$. This makes sense, too, since the event "three fish on the first scoop" must have the same probability as "three fish on the first scoop and no fish on the second." Now consider the third column, corresponding to $X=2$; the total probability in this column is $3\pi^2(1-\pi)=P(X=2)$, and the proportions to be shared between $Y=0$ (bottom cell) and $Y=1$ (second cell) are given by the conditional probabilities $P(Y=0|X=2)$ and $P(Y=1|X=2)$. These have been found to be $1-\pi$ and π, respectively. In order to divide $3\pi^2(1-\pi)$ into proportions $1-\pi$ and π, it is only necessary to multiply by these quantities, giving $P(Y=0, X=2)=3\pi^2(1-\pi)^2$ for the bottom cell in this column, and $P(Y=1, X=2)=3\pi^3(1-\pi)$ for the second cell. These two values have the required sum for total probability $P(X=2)$ and the right proportions for $P(Y=0|X=2)$ and $P(Y=1|X=2)$.

Thus the event "two fish caught on the first scoop and none on the second scoop" has probability $3\pi^2(1-\pi)^2$. The same principle can be applied to get the probabilities in the first two columns, and to fill out the joint probability array as in Table 2.2.[†]

[†] In this book we take up and right as positive directions, exactly as in analytic geometry. Other texts sometimes use down and right.

Table 2.2. Joint probability array

$Y=3$	$\pi^3(1-\pi)^3$	0	0	0
$Y=2$	$3\pi^2(1-\pi)^4$	$3\pi^3(1-\pi)^2$	0	0
$Y=1$	$3\pi(1-\pi)^5$	$6\pi^2(1-\pi)^3$	$3\pi^3(1-\pi)$	0
$Y=0$	$(1-\pi)^6$	$3\pi(1-\pi)^4$	$3\pi^2(1-\pi)^2$	π^3
	$X=0$	$X=1$	$X=2$	$X=3$

This argument about scooping fish represents an entirely typical way of getting from a probability model involving two variables to the complete description, namely the bivariate table. Next consider some of the implications of the table, once obtained.

First, it is possible to get the single-variable distribution of the number of fish caught on the second scoop, $P(Y=y)$, simply by summing the rows:

$$P(Y=0)=(1-\pi)^6+3\pi(1-\pi)^4+3\pi^2(1-\pi)^2+\pi^3=(1-\Upsilon)^3,$$

$$P(Y=1)=3\pi(1-\pi)^5+6\pi^2(1-\pi)^3+3\pi^3(1-\pi)=3(1-\Upsilon)^2\Upsilon,$$

$$P(Y=2)=3\pi^2(1-\pi)^4+3\pi^3(1-\pi)^2=3(1-\Upsilon)\Upsilon^2,$$

$$P(Y=3)=\pi^3(1-\pi)^3=\Upsilon^3,$$

where $\Upsilon=\pi(1-\pi)$.

That is, Y is also a binomial random variable with parameters 3 and $\pi(1-\pi)$. The unconditional distributions of X and Y are often written along the margins of the bivariate table and are known in statistics as "marginal" distributions.

In addition to the bivariate and single-variate distributions, there are also conditional distributions $P(X=x|Y=y)$ corresponding to $P(Y=y|X=x)$. These are formed by normalizing to unity the row or column specified by the condition imposed. When $Y=2$ comes after the vertical bar, it is an instruction to consider row three and normalize to unity:

$$P(X=0|Y=2)=\frac{1-2\pi+\pi^2}{1-\pi+\pi^2},$$

$$P(X=1|Y=2)=\frac{\pi}{1-\pi+\pi^2}.$$

The student should think carefully about these conditional distributions: They give the probability of something which has happened earlier in time (the first scoop) conditional on something which has happened later in time (the second scoop). There is nothing at all peculiar about this. One might say, for example, "of all people who live to age 60, how many took music lessons at age 10" just as well as the converse. However, there are some who find the situation paradoxical, and they are advised to keep thinking about it until the fog lifts.

The various sets of conditional distributions are called in statistics "array" distributions, a row or column being an array. There are therefore five kinds of distributions distinguished so far in a bivariate table:

X	single variable	or	marginal,
Y	single variable	or	marginal,
$X\mid Y$	conditional	or	array,
$Y\mid X$	conditional	or	array,
X, Y	bivariate, joint	or	contingency.

Towards the end of this chapter, some other important kinds of distributions will be discovered in the bivariate table.

Although this example has been given with random variables, it is also true that bivariate distributions apply when the sample points are labels rather than numbers.

The relationships among the various distributions which have been illustrated in this section are now put on a more formal basis in the next section.

2.2. Conditional Probability and Bayes' Theorem

Given two events E and F, the probability of E conditional on F is defined by the formula

$$P(E\mid F)=\frac{P(E\cap F)}{P(F)}. \qquad (1)$$

For the special case of random variable values, the formula reads

$$P(X=x\mid Y=y)=\frac{P(X=x,\ Y=y)}{P(Y=y)}. \qquad (2)$$

The motivation for these definitions should be understood from the fish-netting example: On the left side we have an array distribution, and the right side says that the probabilities in this array distribution are formed by normalizing to unity (the denominator) the joint probabilities (the numerator). Stated yet another way, the "conditional sample space" is F or $Y=y$, and dividing by $P(F)$ or $P(Y=y)$ reflects this fact.

It is also clear that the single-variable probabilities should be defined in terms of array sums of the bivariate probabilities:

$$P(X=x)=\sum_{y} P(X=x,\ Y=y),$$

$$P(Y=y)=\sum_{x} P(X=x,\ Y=y).$$

Since the joint probabilities $P(X=x,\ Y=y)$ are symmetric in the two arguments, the basic definition can be written in the nicely symmetric forms

$$P(E|F)P(F)=P(F|E)P(E) \tag{3}$$

or

$$P(X=x|Y=y)P(Y=y)=P(Y=y|X=x)P(X=x).$$

As a basis for further work, the student should clearly understand why conditional probability *must* be defined in this way, and memorize the simple mnemonic formula

$$\boxed{A|B\cdot B=B|A\cdot A,} \tag{4}$$

from which virtually all more complicated formulas follow easily. It is also useful to see the intuitive significance of these formulas.

Each of the distributions comes with its own system of parameters, distribution functions, probability generating functions, and so forth. There does not seem to be any simple and comprehensive set of notation to account for it all, and rather than attempt to build up such a (cumbersome) system, this book will define the various quantities of interest on an *ad hoc* basis, often using p_{xy} to abbreviate $P(X=x,\ Y=y)$. Thus $p_x=\sum_{y}p_{xy}$, etc.

A problem often involves the basic definition where two of the factors are given, say $P(E)$ and $P(F|E)$ and a third is to be found, usually $P(E|F)$. Solutions clearly involve finding $P(F)$ first. A little reflection will show that this is possible, for the quantities given permit the calculation of the bivariate probabilities $P(E|F)$, and so $P(F)$ is obtained as a row sum.

Example 1. (A Classic Problem). A chest has four drawers, and drawer A contains two gold coins, drawer B contains a silver coin and a gold coin, drawer C contains five silver and two gold coins, drawer D contains two silver and three gold coins. A drawer is chosen at random, and a coin removed—again at random. Being gold, what is the probability that drawer A was opened?

Solution 1. There are two variables, gold/silver and $A/B/C/D$, and therefore a total of eight probabilities in the bivariate table, some of which may be zero. Being given $P(A)=P(B)=P(C)=P(D)=\frac{1}{4}$ ("all drawers equally likely") and the conditionals

$$P(G|A)=1, \qquad P(S|A)=0,$$

$$P(G|B)=\tfrac{1}{2}, \qquad P(S|B)=\tfrac{1}{2},$$

$$P(G|C)=\tfrac{2}{7}, \qquad P(S|C)=\tfrac{5}{7},$$

$$P(G|D)=\tfrac{3}{5}, \qquad P(S|D)=\tfrac{2}{5}$$

("coin chosen at random"), it is perfectly easy to write down the bivariate table and draw from it any conclusions you wish, including the one asked for in the problem. Since $P(A|G)$ was requested, it will be necessary to read off $P(A,G)$ from the table and then divide this by the row sum $P(G)$. Note that in finding $P(G)$ it is necessary to know the joint probabilities, and that to find these it is necessary to know all the conditional probabilities.

There are many problems of this general tenor, which the student should work until the basic procedure is familiar. In the following example, the problem is an industrial one, but the situation, as far as it relates to probability, is virtually identical.

Example 2. Factory A produces 50% of the output, factory B, 25%, and factory C, 25%. In factories A and B, one percent of items are defective, while in factory C three percent are defective. What is the probability that a (randomly chosen) defective item comes from factory A?

Solution 2. $P(A)=\frac{1}{2}$, $P(B)=\frac{1}{4}$, $P(C)=\frac{1}{4}$. $P(D|A)=\frac{1}{100}$, $P(D|B)=\frac{1}{100}$, $P(D|C)=\frac{3}{100}$. The bivariate table is easily constructed and the appropriate value read off.

Problems of this type, in which one collection of conditional probabilities are given, together with one marginal distribution, and the other set of conditional probabilities is to be found, occur so frequently in practice that the steps involved, namely,

1. finding the bivariate probabilities,
2. normalizing to unity the required array,

have been collected into a (famous) formula: Bayes'[†] theorem. To obtain this theorem, let us take the steps in sequence, beginning with being given $P(X=x|Y=y)$ and $P(Y=y)$ and required to find $P(Y=y|X=x)$.

From the basic definition, dividing by $P(X=x)$,

$$P(Y=y|X=x)=\frac{P(Y=y, X=x)}{P(X=x)}.$$

The numerator, being bivariate probabilities, are obtained, as in all earlier examples, by multiplication of the given probabilities:

$$P(Y=y|X=x)=\frac{P(X=x|Y=y)P(Y=y)}{P(X=x)}.$$

The denominator, being a marginal, is obtained by summing the appropriate array:

$$P(Y=y|X=x)=\frac{P(X=x|Y=y)P(Y=y)}{\sum_{y}P(X=x|Y=y)P(Y=y)}. \tag{5}$$

Thus Bayes' theorem is only a slight modification of the basic definition, but it does emphasize one important point: In the symmetric (mnemonic, for example) form, there appear to be four separate distributions. The Bayes' theorem form shows that there are in reality only three, since one of them can be obtained when two are given, for all values of the random

[†] Thomas Bayes, English theologian, 1702–1761.

variable. There is also a version of Bayes' theorem, with the same meaning and the same proof, for events not necessarily involving random variables. For example, in the gold/silver coin problem, Bayes' theorem would read

$$P(A|G) = \frac{P(G|A)P(A)}{P(G|A)P(A) + P(G|B)P(B) + P(G|C)P(C) + P(G|D)P(D)},$$

representing exactly the calculations needed to obtain the bivariate table from the given probabilities, and then the answer from the bivariate table.

It may be surprising how often practical problems involve being given $P(E|F)$ and $P(F)$, rather than some other probabilities; the existence of Bayes' theorem reflects this fact. Although Bayes' theorem permits the calculation of the "inverse" probabilities without writing out the entire bivariate table, it is more sensible and orderly to do the latter, and the actual calculations usually prove to be identical, as the derivation of the theorem shows.

Sometimes the events which condition probabilities are more subtle than obvious random variable values. In the two following examples—both of which will assume importance later in this book—the nature of the conditioning may not at first be clear.

Problem 1 (Stick Problem). There are two sticks, one a foot in length and the other two feet in length. What is the probability distribution for the length of a stick? If a stick is chosen at random, the result $P(X=1)=\frac{1}{2}$, $P(X=2)=\frac{1}{2}$ would be obtained, where $X=$ length of stick. On the other hand, by choosing at random a point on a stick, the result $P(X=1)=\frac{1}{3}$, $P(X=2)=\frac{2}{3}$ would be correct. Actually the difference in the two results comes from an implicit assumption about how the experiment is performed. Some authors call the situation a paradox, although there is certainly no reason to do so. Using the concepts of conditional probability, it appears that the definition of X is not the same in the two cases.

Problem 2 (Family Problem). A family has two children, of which one is a boy. What is the probability that the other is a boy? In this formulation, it appears that the answer must be one-half, since the other child could equally well be boy or girl. But, like the stick problem, there is a curiosity about the way the problem is phrased. It refers exclusively to families with one boy and another child, not to all families. If families with two children have equal probabilities of boys and girls, then the sex distributions which are equally likely would be

boy/boy,   boy/girl,   girl/boy,   girl/girl,

and the problem refers only to the first three kinds of families. Conditional on this, the probability of another boy is one-third. In order to get the answer one-half, it would be necessary to ask the question not in terms of families, but in terms of boys: If a boy has one sibling, what is the probability that it is a boy? Then the two middle cases are indistinguishable, and the answer is one-half.

The student is advised to give some thought to these two examples, since the general principles involved will become more important in Chapter 5.

2.3. Conditioning

The proof of Bayes' theorem shows, among other things, that

$$P(X=x)=\sum_{y} P(X=x|Y=y)P(Y=y), \qquad (6)$$

and this is obvious in any case from the fact that $P(X=x)$ is an array sum. The validity of the formula depends, however, on the fact that all values of y are represented, in other words, that the values are mutually exclusive and exhaustive of the sample space. The same is true of any mutually exclusive and exhaustive set of events, say $E_1, E_2, \ldots, E_n$: If $P(E_i \cap E_j)=0$, all i, j, then

$$P(F)=P(F|E_1)P(E_1)+P(F|E_2)P(E_2)+ \cdots +P(F|E_n)P(E_n). \quad (7)$$

When distributions, probabilities, or expectations are calculated in this way, the procedure is called "conditioning" (on the set of mutually exclusive and exhaustive events); it is often an extremely useful method.

Example 1. A population consists of males and females, who may be colorblind or normal:

$$P(C)=P(C|M)P(M)+P(C|F)P(F).$$

Example 2. A manufactured article is inspected X times, where $X=$ 0,1,2, and has Y flaws:

$$P(Y=y)=P(Y=y|X=0)P(X=0)+P(Y=y|X=1)P(X=1)$$

$$+P(Y=y|X=2)P(X=2).$$

Example 3. In the fish-netting example,

$$P(Y=0)= \sum_{j=0}^{3} P(Y=0|X=j)P(X=j).$$

It is also possible to condition expectation, since it is a linear functional.

Example 4. In the fish-netting experiment, the mean number of fish caught on the second scoop is $3\pi(1-\pi)$, because Y is binomial with parameters 3 and $\pi(1-\pi)$. The same result could be obtained by conditioning on X:

$$E(Y)=E(Y|X=0)P(X=0)+E(Y|X=1)P(X=1)$$

$$+E(Y|X=2)P(X=2)+E(Y|X=3)P(X=3)$$

$$=3\pi(1-\pi)^3+6\pi^2(1-\pi)^2+3\pi^3(1-\pi).$$

Note. The proof of conditioning for expected values depends only on the definition. For random variables defined for non-negative integer values, it would be written

$$E(Y)= \sum_{j=0}^{\infty} jP(Y=j)$$

$$= \sum_{j=0}^{\infty} j \sum_{i=0}^{\infty} P(Y=j|X=i)P(X=i)$$

$$= \sum_{i=0}^{\infty} \sum_{j=0}^{\infty} jP(Y=j|X=i)P(X=i)$$

$$= \sum_{i=0}^{\infty} E(Y|X=i)P(X=i), \tag{8}$$

with similar expressions for other mutually exclusive and exhaustive events. This last formula is often written

$$E(Y)=E(E(Y|X)).$$

This means treating $E(Y|X)$ as a random variable with an expectation. Although the attitude towards $E(Y|X)$ is hardly of importance, it would be well to clarify it and the meaning of the formula by referring once again to the fish-netting example. Treating $E(Y|X=x)$ in this example as values of a random variable 3π, 2π, π, and 0 and probabilities $(1-\pi)^3$, $3\pi(1-\pi)^2$, $3\pi^2(1-\pi)$, and π^3 would lead to exactly the same calculation for $E(Y)$ as performed above.

Example 5. (Pólya's Urn Model for Simple Contagion). Consider a container which holds a amber and b black balls. One ball is drawn at random and replaced in the container, together with x balls of the color drawn. The process is repeated sequentially. What is the probability that the second ball drawn is amber? It is convenient to condition on the first ball drawn:

$$P(\text{second amber}) = P(\text{second amber}|\text{first amber})P(\text{first amber})$$

$$+ P(\text{second amber}|\text{first black})P(\text{first black}).$$

$$= \frac{x+a}{x+a+b}\frac{a}{a+b} + \frac{a}{x+a+b}\frac{b}{a+b}$$

$$= \frac{a}{a+b}.$$

In this example it was fairly obvious that the conditioning should be done on the first drawing, but it can happen that some care must be taken in the conditioning choice. Suppose, to continue, that it is desired to know the probability that the third ball drawn is amber. Use the notation A_n, B_n to abbreviate amber and black on the nth draw. Then, depending on the choice of event to condition on, one of the following formulas might be used:

Conditioning on the Second Draw,

$$P(A_3) = P(A_3|A_2)P(A_2) + P(A_3|B_2)P(B_2);$$

Conditioning on the First Draw,

$$P(A_3) = P(A_3|A_1)P(A_1) + P(A_3|B_1)P(B_1);$$

Conditioning on Both the First and Second Draws,

$$P(A_3) = P(A_3|A_1A_2)P(A_1A_2) + P(A_3|A_1B_2)P(A_1B_2)$$

$$+ P(A_3|A_2B_1)P(A_2B_1)$$

$$+ P(A_3|B_1B_2)P(B_1B_2).$$

The student can work out the details of these formulas.

The important thing to remember, however, is that in each case the events conditioned upon are mutually exclusive and exhaustive.

In terms of the bivariate array, the conditioning formula for the probability of amber on the second draw is easy to understand: It simply states that one of the marginal entries is obtained by summing across the relevant row, and that the entries in that row are obtained as products of conditional and unconditional probabilities. Thus the formulas which have been used are by no means new—they are simply obvious relationships in the bivariate table put to a slightly new purpose.

From the point of view of the substance of the Pólya model, it is interesting to note that the probability of amber on any drawing is the same, whereas the conditional probabilities depend on x. The student should look at a slight variant of this model, in which x balls of the color drawn are added together with y balls of the opposite color, and note that the probability of amber does change as the process continues.

Finally, note that more complicated conditioning can be worked into the basic formula (6), for example,

$$P(X=x|Z=z)=\sum_y P(X=x|Y=y)P(Y=y|Z=z), \qquad (9)$$

a formula which is simultaneously complex and trivial.

2.4. Independence and Bernoulli Trials

The word *independent* (and its opposite, *dependent*) is used in three ways: independent random variables, independent events, and independent experiments (or trials). Random variables are independent if

$$P(X=x, Y=y)=P(X=x)P(Y=y) \qquad \text{for all } x, y, \qquad (10)$$

that is, if the *bivariate table is a multiplication table* of the marginal distributions. Such a decomposition of the joint probabilities is equivalent to

$$P(X=x)=P(X=x|Y=y),$$

$$P(Y=y)=P(Y=y|X=x), \qquad (11)$$

which correspond exactly to our intuitive idea of independence.

Similar formulas define independent events:

$$P(E \cap F) = P(E)P(F),$$

$$P(E) = P(E|F), \tag{12}$$

$$P(F) = P(F|E).$$

Two experiments are independent if the sample points constitute independent events, i.e., if the joint probability of any result of the two experiments can be obtained as the product of the separate probabilities. Independent random variables, events, or experiments constitute an important class of variables, events, and experiments, but not such an important class as is often believed. The importance lies chiefly in statistics, where it is usually necessary (and often desirable) to consider a sample to consist of independent random variables. Actually, much of the interest in probability vanishes with the vanishing of $P(X=x, Y=y)$ as a separate entity; things become "too simple."

Example 1. In the fish-netting experiment, the variables would have become independent if each fish caught were returned to the bowl before the next scoop.

Example 2. In the coin-drawer experiment, the events would have become independent if each drawer contained exactly the same mix.

Example 3. A *Bernoulli experiment* is one with only two possible results, called generically "success" and "failure": on/off, male/female, pass/fail, open/closed, acceptable/defective, graduate/undergraduate, citizen/alien, alive/dead. *Bernoulli trials* are a sequence of independent Bernoulli experiments in which the probability of success $P(S)$ remains constant [with of course $P(F) = 1 - P(S)$]. Thus, for two Bernoulli trials, a bivariate table such as Table 2.3 holds. For n Bernoulli trials, an n-dimensional table similar to Table 2.3 is formed.

Theorem 1. The probability of x successes in n Bernoulli trials, with $P(S) = \pi$, is binomial with parameters n and π.

Proof. The probability of x successes on the first x trials is π^x; the probability of $n-x$ failures on the next $n-x$ trials is $(1-\pi)^{n-x}$. Since these

Table 2.3. Bivariate table for two Bernoulli trials

Second	$P(F)$	$P(S)P(F)$	$P(F)P(F)$
trial	$P(S)$	$P(S)P(S)$	$P(F)P(S)$
		First trial	
	$P(S)$		$P(F)$

may happen in any order (and with the same probability), the factor $\binom{n}{x}$ is supplied.   □

Theorem 2. The number of failures before n successes in Bernoulli trials is negative binomial with parameters n and π.

Proof. Let Y denote the number of failures before n successes. The event $Y=y$ will occur when the nth success occurs on the $(n+y)$th trial. Thus $P(Y=y)$ is the product of the probability of y failures in the first $n+y-1$ trials, and the probability of a success on the $(n+y)$th trial, namely

$$\binom{n+y-1}{y}(1-\pi)^y \pi^{n-1}\pi, \qquad y=0,1,2,\ldots,$$

as given in Section 1.10.   □

The binomial and negative binomial distributions often appear in this interpretation based on random variables defined in connection with Bernoulli trials. The student should beware, however, of thinking that this is the sole interpretation of the two distributions. They also occur in many other contexts.

The Method of Marks. The interpretation of the binomial distribution as the probability of x successes in n independent trials permits an ingenious derivation of the binomial based on an idea known as the method of marks. Suppose each success is "marked"—you can think of red paint—with probability $1-s$, and left unmarked with probability s, each success being treated independently, and all failures left unmarked. Then the probability that none of the trials is marked is given by

$$\sum_{j=0}^{n} P(X=j)s^j,$$

where, as before, X denotes the number of successes. But this probability can also be written $(1-\pi+\pi s)^n$, since $1-\pi$ is the probability of a failure, which is never marked, and πs is the probability of an unmarked success.

Equating these expressions for the probability of no marks leads to the probability generating function of the binomial distribution.

This method, although admittedly peculiar, does (a) apply to more complicated problems, as will be shown later in the book, and (b) provide an interpretation, also peculiar, for the variable s in the probability generating function.

Binomial–Poisson Limit. Consider a line segment of length ℓ, contained within a longer segment of length L. Suppose points are dropped at random on the longer segment so that the probability of a given point falling on ℓ is ℓ/L. Placing n points on L can be modeled as a Bernoulli experiment, with $\pi = \ell/L$, a "success" being a point falling on ℓ. The probability that x points will fall on ℓ is therefore binomial with parameters n and ℓ/L.

Now suppose that the number n of points is increased and the length of L is also increased, so that both $\to \infty$, but proportionally, so that $n/L \to \lambda$, a constant representing the long-run density of points on the line. Then, as the following calculation shows, the probability of x points on ℓ, originally binomial, now approaches the Poisson with parameter $\ell\lambda$:

$$\lim \binom{n}{x}\left(\frac{\ell}{L}\right)^{x}\left(1-\frac{\ell}{L}\right)^{n-x} = \frac{\ell^{x}}{x!}\lim\left(\frac{n}{L}\right)\left(\frac{n-1}{L}\right)\cdots\left(\frac{n-x+1}{L}\right)\left(1-\frac{\ell}{L}\right)^{L\lambda-x}$$

$$= \frac{(\ell\lambda)^{x}e^{-\ell\lambda}}{x!}.$$

It will be noted that the exponential limit

$$e^{z} = \lim_{n \to \infty}\left(1+\frac{z}{n}\right)^{n}$$

is used in this proof.

This limit has an important practical interpretation. Suppose the line is the time axis (with $L = \infty$) and the points are instants of occurrence of some well-defined phenomenon, such as accidents in a factory, lightning bolts in a county, registrations on a Geiger counter, demands for service at a counter, calls placed at a switchboard, etc. Then, the binomial–Poisson limit shows that if the instants are random in time, that is, completely independent and scrambled as to when the phenomena occur, the Poisson distribution gives the probability of x occurrences in a fixed unit time, where the parameter is the average number of occurrences in a unit time.

   Another interpretation of the limit, often used in statistics, is to show that if n is large and π is small, the binomial probabilities can be approximated by the Poisson probabilities. It often turns out that the Poisson probabilities (with only one parameter) are a bit easier to calculate than the binomial probabilities (with two parameters).

2.5. Moments, Distribution Functions, and Generating Functions

   Each of the five kinds of distributions which have so far been discussed in connection with two random variables carries with it the entire structure of moments, distribution functions, and generating functions defined in Chapter 1. In thinking about the various quantities and functions involved even when there are two random variables, let alone n random variables, it appears that a very complex set of notation would be needed to describe everything systematically. It is probably safe to say that there does not exist any such fully comprehensive set, and if one were devised, the tangle of symbols would obscure rather than reveal important relationships.

   In this section, some of the more useful relationships are illustrated, together with typical notation. In the remainder of the book, various quantities and functions will be defined according to the need to clarify the problem at hand, without attempting an overall system.

   Means:

$$E(X)=\sum_i \sum_j iP(X=i, Y=j),$$

$$E(X|Y=y)=\sum_i iP(X=i|Y=y)=\frac{\sum_i iP(X=i, Y=y)}{P(Y=y)},$$

etc.

   Variances:

$$\mathrm{var}(X)=\sum_i \sum_j i^2 P(X=i, Y=j)-\left[E(X)\right]^2,$$

$$\mathrm{var}(X|Y=y)=\sum_i i^2 P(X=i|Y=y)-\left[E(X|Y=y)\right]^2,$$

etc.

With two random variables, three closely related quantities are often used to measure the degree of association between the variables:

$$E(XY)=\sum_i \sum_j ijP(X=i,Y=j),$$

$$\mathrm{cov}(X,Y)=E(XY)-E(X)E(Y) \qquad \text{(the "covariance")}, \qquad (13)$$

$$\rho(X,Y)=\mathrm{cov}(X,Y)/[\mathrm{var}(X)\mathrm{var}(Y)]^{1/2} \qquad \text{(the "correlation coefficient")}.$$

When

$$E(XY)=E(X)E(Y),$$

the variables are said to be uncorrelated. The student can verify that independent variables are uncorrelated, and perhaps construct a counterexample to show that the converse is not true.

Probability Generating Function

Defining

$$\phi(s,t)=\sum_i \sum_j s^i t^j P(X=i,Y=j) \qquad (14)$$

for the bivariate distribution and

$$\alpha(s)=\sum s^i P(X=i),$$

$$\beta(s)=\sum s^i P(Y=i)$$

for the single-variable marginals, it is obvious that $\phi(s,1)=\alpha(s)$ and $\phi(1,t)=\beta(t)$. Also,

$$\left.\frac{\partial\phi(s,t)}{\partial s}\right|_{s=t=1}=E(X)$$

and

$$\left.\frac{\partial^2(s,t)}{\partial s\,\partial t}\right|_{s=t=1}=E(XY). \qquad (15)$$

It is also possible to define the probability generating functions for the conditional distributions in the usual way, with the usual set of relationships following.

Distribution Functions

Defining the bivariate distribution function by

$$F(x, y) = P(X < x, Y < y)$$

and the single-variable distribution functions by

$$G(x) = P(X < x),$$

$$H(x) = P(Y < x),$$

the relationships $F(x, \infty) = G(x)$, $F(\infty, x) = H(x)$ follow at once.

In contrast to the single-variable distribution function, however,

$$1 - F(x, y) \neq P(X \geq x, Y \geq y),$$

since the events $(X < x) \cap (Y < y)$ and $(X \geq x) \cap (Y \geq y)$ do not exhaust the sample space.

Notation. Some authors prefer to keep the same letter for all distribution functions, indicating the random variable as a subscript, thus: $F_{X,Y}(x, y)$, $F_X(x)$, $F_Y(y)$, and so forth. This system is also often used for other functions, such as the probability generating function and even the basic exact probabilities, as noted in Section 1.4. As with most notational issues, there are advantages and disadvantages to any system, and both seem to sharpen as the number of variables increases. Fortunately, the problems addressed in this book require only *ad hoc* notation.

2.6. Convolutions and Sums of Random Variables

In the fishing example of Section 2.1, let $Z = X + Y$ be the number of fish caught in the first two trials, possible values being 0, 1, 2, and 3. Just as the probabilities for values of X are found in the horizontal sums (marginal for X) and values of Y for the vertical sums, so the probabilities for values of Z are found in diagonal sums. The value $Z = 0$ occurs only when both $X = 0$ and $Y = 0$, i.e., the probability in the lower left corner, $(1 - \pi)^6$. To have $Z=1$, either of the events $X=0$, $Y=1$ or $X=1$, $Y=0$ must occur, and these probabilities are in the next diagonal:

$$P(Z=1)=3\pi(1-\pi)^5+3\pi(1-\pi)^4.$$

Similarly,

$$P(Z=2)=3\pi^2(1-\pi)^4+6\pi^2(1-\pi)^3+3\pi^2(1-\pi)^2$$

and

$$P(Z=3)=\pi^3(1-\pi)^3+3\pi^3(1-\pi)^2+3\pi^3(1-\pi)+\pi^3.$$

The sum of all probabilities equals unity simply because it is composed of sums of all the values in the bivariate table.

In general, if

$$p_{xy}=P(X=x,Y=y),$$

then

$$P(X+Y=z)=\sum_{j=0}^{z}p_{j,z-j}, \tag{16}$$

where the values in the summation depend on the possible values of X and Y. However, even if X and Y have an infinite number of values, the sums (being diagonals) are each still finite, although there are an infinite number of such sums.

Once the significance of the diagonal sums as probabilities is appreciated, the fact can be used in other ways, for example, in constructing the bivariate table.

Example 1. In a population where males and females occur with equal probability, the number of people entering a shop during a day is Poisson with parameter λ. Find the probability that on one day, x males and y females enter the shop. Here the Poisson probability $e^{-\lambda}\lambda^z/z!$ is given as the sum of the probabilities on the $(z+1)$st diagonal. Within the diagonal, the proportions of different combinations of males and females is binomial distributed with parameters z and $\frac{1}{2}$. Thus

$$P(X=x, Y=y) = \frac{e^{-\lambda}\lambda^{x+y}}{(x+y)!}\binom{x+y}{x}2^{-x-y}$$

(see Problems 65 and 66).

In this example, it is clear from the problem (or from finding the marginals and multiplying) that the random variables are independent, so that formula (16) can be written

$$P(X+Y=z) = \sum_{j=0}^{z} p_j q_{z-j}, \tag{17}$$

where $P(X=x)=p_x$ and $P(Y=x)=q_x$.

Relationship (17) is an important one in mathematics and is called the *convolution* of the sequences p_x and q_x. The general relation, written with an asterisk,

$$\{a\} * \{b\} = \{c\},$$

denotes a relationship between sequences $a_0, a_1, a_2, \ldots,$ $b_0, b_1, b_2, \ldots,$ $c_0, c_1, c_2, \ldots,$ where

$$c_0 = a_0 b_0,$$

$$c_1 = a_0 b_1 + a_1 b_0,$$

$$c_2 = a_0 b_2 + a_1 b_1 + a_2 b_1,$$

$$c_3 = a_0 b_3 + a_1 b_2 + a_2 b_1 + a_3 b_0,$$

$$\vdots \qquad \qquad \vdots$$

exactly as specified by Eq. (17) for probabilities.

When random variables are independent, the probability generating function of the sum is equal to the product of the probability generating functions. This can be verified by multiplying the two series together:

$$\left(p_0 + sp_1 + s^2 p_2 + s^3 p_3 + \cdots\right)\left(q_0 + sq_1 + s^2 q_2 + s^3 q_3 + \cdots\right)$$

yields a power series in which the coefficient of s^x is the xth sum in the convolution of p and q. Thus, for the case of independence,

Random Variables: Addition;

Distribution: Convolution;

Probability generating function: Multiplication.

This shows, incidently, that convolutions cannot be considered to have "inverses"; they should not be regarded in any way as a kind of multiplication, because the probability generating function for a single throw of a fair die can be "decomposed" into probability generating functions in two different ways:

$$\tfrac{1}{6}s(1+s+s^2+s^3+s^4+s^5)=\left[\tfrac{1}{3}(1+s+s^2)(\tfrac{1}{2}s(1+s^3))\right]$$

$$=\left[\tfrac{1}{3}(1+s^2+s^4)(\tfrac{1}{2}s(1+s))\right].$$

The relationship between probability generating functions corresponding to sums of independent random variables can also be obtained by a direct method. Suppose X and Y are independent random variables with probabilities $p_x=P(X=x)$ and $q_y=P(Y=y)$ and probability generating functions $\phi(s)$ and $\psi(s)$, respectively. Consider X Bernoulli trials with a probability s of success on a given trial. Then the probability of no failures will be $\phi(s)$. Similarly, in Y Bernoulli trials with probability s of failure, the probability of no failures will be $\psi(s)$. The probability of no failures on $X+Y$ consecutive trials is therefore $\phi(s)\psi(s)$, which, according to the argument, is the probability generating function of the random variable $X+Y$.

In discussing sums of random variables, the emphasis on convolutions (case of independence) should not obscure the fact that exactly the same principle (diagonal sums in the bivariate table) holds when the variables are not independent. If there is any special simplicity in the independence case, it lies in obtaining the joint distribution, not in summing the diagonals.

Moments of Sums of Random Variables. There are a number of established formulas relating to the moments of the distribution of $X+Y$, most of which are not needed in this book. However, the student should note—and confirm by proof—the following simple ones:

$$E(X+Y)=E(X)+E(Y),\tag{18}$$

$$\text{var}(X+Y)=\text{var}(X)+\text{var}(Y)+2\text{cov}(X,Y).$$

When the variables are independent, the last term vanishes.

Finally, for two constants a and b, consider the expression

$$E\big((aX+bY)^2\big)=a^2E(X^2)+2abE(XY)+b^2E(Y^2).$$

Since the left side of the identity is positive for all values of a and b, so must the right side be always positive. This implies that zeros of the quadratic form must be complex, and therefore the discriminant must be negative:

$$\big[2E(XY)\big]^2-4E(X^2)E(Y^2)\leq0,$$

which leads to the well known Cauchy–Schwarz inequality

$$\big[E(XY)\big]^2\leq E(X^2)E(Y^2).$$

This formula is sometimes useful in statistics. Naturally, all the formulas relating to moments depend on having a distribution for which the appropriate moments exist (i.e., converge).

2.7. Computing Convolutions: Examples

The distribution of the sum of two independent random variables can be obtained from the separate distributions either directly or by using the probability generating functions. In this section, some examples are given for various simple distributions.

Poisson Distribution. If X and Y are independent Poisson random variables with probability generating functions $\exp(-\lambda+\lambda s)$ and $\exp(-\mu+\mu s)$, then it is obvious that $X+Y$ is also a Poisson random variable, since the product of the probability generating functions is of the same form, with parameter $\lambda+\mu$.

Geometric Distribution. If X and Y are independent geometric random variables with the same parameter ρ, then the sum $X+Y$ has probability generating function $(1-\rho)^2/(1-\rho s)^2$, which is of the form $(1-\pi+\pi s)^{-2}$, with $(1-\rho)(1-\pi)=1$. Thus $X+Y$ has the negative binomial distribution with parameters 2 and π. When the independent geometric random variables do not have the same parameter, the distribution of the sum is much more complicated, as can be seen by expanding the reciprocal of a general quadratic.

In order to obtain the Poisson result directly by convolutions, it would be necessary to sum

$$P(X+Y=z)= \sum_{j=0}^{z} P(X=j)P(Y=z-j)$$

$$= \sum_{j=0}^{z} \frac{e^{-\lambda}\lambda^j}{j!} \frac{e^{-\mu}\mu^{z-j}}{(z-j)!}.$$

This is essentially a binomial sum yielding

$$\frac{e^{-(\lambda+\mu)}(\lambda+\mu)^z}{z!}.$$

In the asterisk notation, this fact could be written

$$\frac{e^{-\lambda}\lambda^x}{x!} * \frac{e^{-\mu}\mu^x}{x!} = \frac{e^{-(\lambda+\mu)}(\lambda+\mu)^x}{x!}$$

or, when $\lambda=\mu$,

$$\left(\frac{e^{-\lambda}\lambda^x}{x!}\right)^{2*} = \frac{e^{-2\lambda}(2\lambda)^x}{x!},$$

with the obvious extension to $n*$. With this system, the derivation of the binomial distribution from the Bernoulli could be written

$$\left[\pi^x(1-\pi)^{1-x}\right]^{n*} = \binom{n}{x}\pi^x(1-\pi)^{n-x},$$

and of the negative binomial from the geometric,

$$\left[(1-\rho)\rho^x\right]^{n*} = \binom{n+x-1}{x}\rho^x(1-\rho)^n.$$

When a convolution is performed $n-1$ times—that is, involving n distributions—the result is said to be an *n-fold* convolution. Also, it is convenient to define the zero-fold convolution of any distribution to be the causal distribution at the origin:

$$(p_x)^{0*} = \delta(x) = \begin{cases} 1, & x=0, \\ 0, & x \neq 0, \end{cases}$$

and

$$(p_x)^{1*} = p_x.$$

Convolutions can also be applied to the distribution functions, but *not* with the result that $F * F$ is the distribution function of the sum of the random variables. In fact, to obtain the distribution function of the sum, it is necessary to apply the convolution to the distribution function of one independent variable and the exact probabilities of the other, as shown by the following proof:

$$P(X+Y<x) = \text{the sum of the probabilities in a}$$
$$\text{triangle } (0,0), (0, x-1), (x-1,0)$$
$$\text{in the bivariate table}$$

$$= \sum_{j=0}^{x-1} \sum_{i=0}^{x-j-1} p_i q_j,$$

where $P(X=x)=p_x$, $P(Y=y)=q_y$. Changing variable j to k, where $x-j=k$,

$$P(X+Y<x) = \sum_{k=1}^{x} \sum_{i=0}^{k-1} p_i q_{x-k}$$

$$= \sum_{k=1}^{x} P(X<k)P(Y=x-k).$$

2.8. Diagonal Distributions

In Section 2.6, distributions were formed by considering the sum of probabilities on a diagonal to be a new probability; in this section a diagonal is chosen, and then its elements normalized to unity by dividing by the diagonal sum. These distributions are conditional on the sum of the two random variables being fixed.

For example, in the familiar fish-netting experiment, suppose the total number of fish caught in both scoops is two. Then with a little calculation, the new distribution is obtained:

$$P(X=0|X+Y=2)=P(Y=2|X+Y=2)=\frac{(1-\pi)^2}{(2-\pi)^2},$$

$$P(X=1|X+Y=2)=P(Y=1|X+Y=2)=\frac{2(1-\pi)}{(2-\pi)^2},$$

$$P(X=2|X+Y=2)=P(Y=0|X+Y=2)=\frac{1}{(2-\pi)^2}.$$

The student should verify this result and find the other diagonal distributions for this example, with the general result

$$P(X=x|X+Y=n)=\binom{n}{x}\beta^x(1-\beta)^{n-x},$$

$$x=0,1,\ldots,n,\quad n=0,1,2,3,$$

where

$$\beta=\frac{1}{2-\pi}.$$

There are a number of interesting relationships involving diagonal distributions, as well as some important applications.

Inasmuch as a number of different distributions are involved, a few abbreviations will be introduced:

$$a_x=P(X=x),\qquad b_x=P(Y=x)$$

for the marginal distributions. No assumption of independence is being made, so a special symbol is needed for the joint probabilities:

$$p_{xy}=P(X=x,\ Y=y).$$

The diagonal sums will be denoted by

$$c_x=P(X+Y=x),$$

so that the probabilities in the nth diagonal, defined by

$$q_n(x) = P(X = x | X + Y = n),$$

satisfy

$$c_n q_x(n) = p_{x, n-x}, \qquad (19)$$

which can be written

$$c_{x+y} q_x(x+y) = p_{xy},$$

thus expressing the joint probability distribution in terms of the diagonals. The same relationship can be written in probability generating functions

$$\alpha(s) = \sum a_j s^j, \qquad \beta(s) = \sum b_j s^j,$$

$$\phi(s, t) = \sum \sum p_{ij} s^i t^j, \qquad \psi_n(s) = \sum q_j(n) s^j$$

as follows:

$$\phi(s, t) = \sum_x \sum_y q_x(x+y) c_{x+y} s^x t^y$$

$$= \sum_{k=0}^{\infty} \sum_{x=0}^{k} q_x(k) c_k s^x t^{k-x}$$

$$= \sum_{k=0}^{\infty} \psi_k(s/t) c_k t^k.$$

Thus the marginal probability generating functions can be written

$$\alpha(s) = \phi(s, 1) = \sum_{k=0}^{\infty} \psi_k(s) c_k, \qquad (20)$$

$$\beta(s) = \phi(1, s) = \sum_{k=0}^{\infty} \psi_k(1/s) s^k c_k. \qquad (21)$$

When the random variables X and Y are independent, the distributions on the diagonals actually determine the marginals. Setting $n = x(=k)$ in Eq. (19) and substituting the value for c_k into Eq. (20) gives

$$\alpha(s) = \sum_{k=0}^{\infty} \psi_k(s) \frac{a_k b_0}{q_k(k)}. \tag{22}$$

The value of b_0 can be found by setting $s = 1$ and then Eq. (22) becomes

$$\alpha(s) = \frac{\sum\limits_{k=0}^{\infty} \psi_k(s)[a_k/q_k(k)]}{\sum\limits_{k=0}^{\infty} [a_k/q_k(k)]}. \tag{23}$$

For example, if the diagonal distribution is binomial, $q_k(k) = \pi^k$, $\psi_k(s) = (1 - \pi + \pi s)^k$ and hence

$$\alpha(s)\alpha(1/\pi) = \alpha[(1 - \pi + \pi s)/\pi], \tag{24}$$

and it will be easily confirmed that this functional equation is satisfied by the Poisson probability generating function $\alpha(s) = \exp(-\lambda + \lambda s)$ for any value of λ.[†] Similarly, $\beta(s) = \exp(-\mu + \mu s)$, subject to the condition that

$$\frac{\lambda}{\mu} = \frac{\pi}{1 - \pi}. \tag{25}$$

It is left as an exercise for the student to show that the converse is true. Diagonal distributions for independent Poisson variables are binomial, with parameter satisfying Eq. (25).

When the diagonal distributions are rectangular (and the random variables are independent),

$$\psi_k(s) = \frac{1 - s^{k+1}}{(1+k)(1-s)}$$

and $s^k \psi_k(1/s) = \psi_k(s)$, showing, from Eq. (21), that $\alpha(s) = \beta(s)$, so that X

[†] It is possible to prove that only this value can be obtained from the functional equation (24); see, for example, ACZEL, J. (1966), *Lectures on Functional Equations and Their Applications*, Academic Press, New York, p. 67 (Theorem 2).

and Y are equidistributed. In fact, Eq. (20) becomes

$$\alpha(s) = \frac{1 - s\alpha(s)}{[1 + E(X)](1 - s)},$$

which corresponds to a geometric distribution with parameter

$$\rho = \frac{E(X)}{1 + E(X)}.$$

It must be emphasized that these results depend on the assumption of the independence of X and Y; indeed the fish-netting example shows a binomial diagonal distribution where the marginals are not Poisson.

The most important application of diagonal distributions occurs in modeling "before-and-after" studies, especially in connection with the occurrence of accidents.[†] Let X be the number of accidents before a supposed improvement is made, and Y, the number of accidents afterwards. If the improvement affected hundreds or thousands of road segments with greatly differing usage characteristics, it might be desirable to "standardize" the analysis by considering together all those segments for which $X + Y$ is fixed. Then the distributions of X and Y separately would enable the experimenter to draw conclusions regarding the effectiveness of the supposed improvement. Furthermore, this could be done without assuming that X and Y are independent.

Recommendations for Further Study

The classic textbook of Feller (1968), although originally published thirty years ago, retains a preeminent position among books on discrete probability. Combining a wealth of material with a clear style and organization, it is surely even now required reading for serious students. An alternative would be the meticulously precise introduction by Chung (1974), especially for those primarily interested in the mathematical structure of probability. At a somewhat more advanced level, the book by Gnedenko (1968) will also be accessible to those having successfully completed the first two chapters of this book.

[†] This model has been used in practical analysis of accident data: see ERLANDER, S. (1971), A review of some statistical models used in automobile insurance and in road accident studies, *Accident Analysis and Prevention*, Vol. 3, No. 1, pp. 45–75, and especially the comprehensive reference list.

CHUNG, KAI LAI (1974), *Elementary Probability Theory with Stochastic Processes*, Springer-Verlag, New York..

FELLER, WILLIAM (1968), *An Introduction to Probability Theory and Its Applications*, Vol. I, third edition, John Wiley and Sons, New York.

GNEDENKO, B. V. (1968), *The Theory of Probability*, fourth edition, Chelsea, New York.

2.9. Problems

Unless otherwise specified, it is assumed in this section that the dice, coins, selection of balls, etc., are unbiased, that is, that each possible result has the same probability. The first 23 problems give practice in setting up a bivariate probability table. The same problems can be used at the instructor's discretion to find properties of these distributions: expected values and other moments, covariance and correlation, generating functions. These can be obtained for marginals, conditional distributions, diagonal distributions of both types, and also various probability questions such as $P(X|Y)$, $P(XY \text{ odd})$, $E(\max(X, Y))$, $\text{var}(X|Y=y)$, and so forth. Independence can be investigated. For extra variety, and simplicity, dice problems can be interpreted as referring to tetrahedral dice, with sides 1, 2, 3, 4. Some of the problems can be made easier by specifying a value of n, others more difficult by replacing a given integer by n.

1. A coin is tossed three times; $X=0$ or 1, according to whether a head or tail occurs on the first throw, and $Y=$ number of heads.

2. Two dice are thrown; $X=$ sum, $Y=$ larger value.

3. Two dice are thrown; $X=$ sum, $Y=$ absolute value of difference.

4. A die is thrown; $X=$ twice the number appearing, $Y=1$ or 3, according to whether the result is odd or even.

5. A coin is tossed four times; $X=$ number of heads, $Y=$ number of runs, i.e., consecutive sequences of heads or tails.

6. Two dice are thrown; $X=$ sum. One die is left in place and the other thrown again; $Y=$ new sum.

7. A card is chosen at random from n consecutively numbered cards; $X=$ number shown. A second card is chosen from the cards numbered $1, 2, \ldots, X$ and its number is Y.

8. A coin is thrown three times; $X=$ number of heads on first two throws, $Y=$ number of tails on last two throws.

9. A die is successively rolled; $X=$ number of rolls needed to obtain a five, $Y=$ number of rolls needed to obtain a six.

10. Two balls are selected from a box with balls labeled 1, 2, 3; $X=$number on first ball selected, $Y=$number on second ball selected (no replacement, of course).

11. A red and a blue die are thrown, with $X=$score on the blue die, and $Y=$larger score.

12. X, Y are as in the fish-netting experiment, but with the basic distribution rectangular rather than binomial.

13. Given the fish-netting experiment with n fish and m scoops, show that the expected number of fish caught decreases with each successive scoop.

14. Three dice are thrown together, the highest number X noted, and that die put to one side. (If two or more show the same number, the throw is invalid, and the experiment repeated.) The remaining dice are thrown until they show different numbers, and the larger number is Y.

15. X has probability generating function $\frac{1}{6}(2s+1)(s+1)$, Y conditional on $X=x$ is rectangular with values x, $x+1$, $x+2$.

16. Cubical dice are painted in an unorthodox manner: 1, 1, 1, 3, 3, 5 on the sides. Two are thrown, with $X=$sum, $Y=$lesser value.

17. A coin is flipped repeatedly; $X=$number of flips needed to get the first head, $Y=$number of flips needed to get the first tail.

18. Three red and two green balls are in a bag. They are taken out one by one until both green balls have been chosen. $X=$number of balls before the first green one, $Y=$number of additional balls before the second green one.

19. Let $P(U=x)=(1-\rho)\rho^x$, $x=0,1,2,\ldots$, and let n be a positive integer. $X=\max(U,n)$, $Y=\min(U,n)$.

20. A die is rolled; $X=$number on top, $Y=$number on side most nearly facing you. (Note: The dice are painted so that the total of the numbers on opposite faces is seven.)

21. Balls numbered 0, 1, 1, 3, 3, 6 are placed in a bag and two are drawn; $X=$larger number, $Y=$total.

22. A bag contains three white and two red balls. Two are drawn consecutively, with $X=$number of white balls on first draw, $Y=$number of white balls on second draw.

23. $\Sigma P(X=x)s^x=\exp(-\lambda+\lambda s)$, $P(Y=y\,|\,X=x)=(1+x)^{-1}$, $y=0,1,2,\ldots,x$.

24. Two (unequal) digits are taken from the set $1,2,\ldots,9$, with every pair having an equal chance of being selected. (i) If the sum is odd, what is the probability that one of the digits is 2? (ii) If one of the digits is 2, what is the probability that the sum is odd?   *Ans*: (i) $\frac{5}{20}$, (ii) $\frac{5}{8}$

25. There are N coins in a box, of which n are normal and $N-n$ are double-headed. A coin is selected at random and tossed r times, each time coming up heads. What is the probability that it is normal?   *Ans.* $n/[n+N(N-n)2^n]$

26. In a certain factory, machine A produces p percent of the output, and x percent of its production is defective. Machine B produces $100-p$ percent of the output, and y percent of its production is defective. What is the probability that a randomly selected defective item came from machine A?

27. Box A contains three green and two yellow balls; box B contains one yellow and two green balls; box C contains one green and three yellow balls. (i) One box is selected at random and a ball drawn at random from it. What is the probability that the ball drawn is yellow? (ii) If a yellow ball is obtained, what is the probability that it came from box C?   *Ans.* (i) $\frac{89}{180}$, (ii) $\frac{45}{89}$

28. Die A has four red and two white faces, whereas die B has two red and four white faces. A fair coin is flipped; if it falls heads, the game continues by throwing die A alone; if it falls tails, die B is used. (i) Show that the probability of red at any throw is $\frac{1}{2}$. (ii) If the first two throws result in red, what is the probability of red at the third throw? (iii) If red turns up at the first n throws, what is the probability that die A is being used?

29. X is binomial with parameters n and π; Y conditional on the value $X=x$ is binomial with parameters x and σ.

30. In searching for a penny, it is known that it is hidden in one of three places, and it is equally likely to be in any one of them. The probability of finding the penny if it is in place x ($x=1,2,3$) is p_x. Suppose a search is made in place one, and the penny is not found. What is the probability that it was there?

31. Twice as many women use the library as men, but of those using the library, women check out an average of two books per visit and men, an average of three. (i) Find the average number of books checked out by a library user. (ii) If a person checks out three books, find the probability that that person is a man, assuming that the number of books checked out (for either sex) is geometric with positive integral values.   *Ans.* (i) $\frac{7}{3}$, (ii) $\frac{16}{43}$

32. Five boxes contain black/white balls as follows: Box A, 2/2; box B, 3/1; box C, 3/2; box D, 0/1 box E, 1/2. If a ball is chosen from each box, what is the expected number of white balls obtained? If a box is selected and then a ball chosen, what is the probability that it is white? If a black ball is obtained, what is the probability that it came from box 4?

33. Consider the formula $P(E|F)P(F)=P(F|E)P(E)$ for a two-by-two table: E, E^c, F, F^c. Bayes' theorem shows that it is sufficient to be given $P(E|F)$, $P(F)$, and $P(E|F^c)$ to determine the whole table. Find counterexamples to show that it is not sufficient to have one factor from each side of the equation given.

34. Box A contains one red and one yellow ball; box B contains one yellow and two red balls. A box is chosen and a yellow ball extracted; what is the probability that it was from box A?

35. Suppose there are four boxes, labeled A, B, C, and D. A ball is chosen at random from box A, which contains initially six balls labeled B, three labeled C, and three labeled D. The letter drawn tells which box is to be used for the

second drawing. Box B contains five green and five yellow balls, box C contains four green and six yellow balls, and box D contains two green and eight yellow balls. Given that the second ball drawn is green, what is the probability that box B is being used? Are the events "first ball C" and "second ball yellow" independent?

36. A fair die is thrown until two consecutive results are the same. What is the expected number of throws?

37. A fair coin is tossed until the same side appears twice in succession, so that $1/2^{n-1}$ is the probability of every result that requires n tosses. Let E be the event that the experiment ends before the sixth toss, and let F be the event that the experiment ends on an even number of tosses. (i) Find $P(E)$ and $P(F)$. (ii) Show that E and F are independent. (iii) If G is the event that the experiment ends before the fifth toss, are F and G independent?

38. In an investigation of animal behavior, rats have to choose between four similar doors, one of which is "correct." If an incorrect choice is made, the rat is returned to the starting point, and made to choose again, this continuing until the correct response is made. The random variable X is the trial on which a correct response is first made, with possible values 1, 2, 3,.... Find $P(X=x)$ and $E(x)$ under the following hypotheses. (i) Each door is equally likely to be chosen on each trial and all trials are mutually independent. (ii) At each trial, the rat chooses with equal probability between doors which have not yet been tried, no choice ever being repeated. (iii) The rat never chooses the same door on two successive trials, but otherwise chooses at random with equal probabilities.
 Ans. Expected values: (i) 4, (ii) $\frac{30}{12}$, (iii) $\frac{13}{4}$

39. Suppose X balls are distributed at random into n boxes, where X is Poisson (over the non-negative integers) with parameter λ. Let Y be the number of empty boxes. Show that Y is binomial with parameters n and $e^{-\lambda/n}$.

40. Two dice are thrown n times. Show that the number of throws in which the number on the first die exceeds the number on the second die is binomially distributed with parameters n and $\frac{5}{12}$.

41. A pair of coins is thrown. (i) What is the distribution of the number of throws needed for both coins to show heads? (iii) What is the distribution of the number of throws needed for at least one coin to show a head?

42. Let X be the total showing on a single throw of n dice. Find $E(X)$ and var(X) as a function of n.

43. In a sequence of Bernoulli trials with $p=P(S)$, let p_x be the probability that the combination SF occurs for the first time on trials number $x-1$ and x. Find the generating function, the mean, and the variance.   *Ans.* Mean$=[p(1-p)]^{-1}$

44. Two players try alternately to obtain a success in a game where the probability of success is p. Show that the probability that the first player wins is $(2-p)^{-1}$. Generalize to n players.

45. A game between two players, A and B, consists in them taking turns playing a machine until one of them scores a success. The first to score a success is the winner. Their probabilities of success in a single play are p for A and q for B. Since B is the better player ($q>p$), he allows A to have the first turn. All plays are independent. (i) Show that the game is fair if and only if $q=p/(1-p)$. (ii) Show that if A wins, the average number of plays he takes in which to win is $(p+q-pq)^{-1}$.

In Problems 46–54 inclusive, it is assumed that X and Y are independent. Where values are not specified, it is also assumed that the Poisson and geometric distributions are defined over the non-negative integers.

46. Both X and Y are Poisson λ; find $P(X=x|X+Y=n)$.

47. If X is Poisson λ and Y is Poisson μ, find $E(X|X+Y=n)$.

48. $P(X=x)=(1-\rho)\rho^{x-2}$, $x=2,3,4,\ldots$, $P(Y=y)=\frac{1}{3}$, $y=1,2,3$. Find $P(X+Y=x)$.

49. Both X and Y are rectangular over $1,2,\ldots,n$. Find $P(X<Y)$, $P(X=Y)$, the distribution of $\min(X,Y)$, and the distribution of $|X-Y|$.

50. X is rectangular over $1,2,3$; Y is rectangular over $1,2,3,4$. Find (i) $P(X=Y)$, (ii) $E(\max(X,Y))$, (iii) $P(XY=x)$.   Ans. (i) $\frac{1}{4}$, (ii) $\frac{17}{6}$

51. For X geometric with parameter ρ and Y geometric with parameter σ, find (i) $P(X+Y=x)$, (ii) $P(X<Y)$, (iii) $P(X=x|X+Y=n)$.

52. Let X, Y be geometric with the same parameter ρ. Let $U=\min(X,Y)$, $V=\max(X,Y)$. (i) Show that

$$P(U=x,V=x+y)=2(1-\rho)^{2x+y}\rho^2, \qquad x=1,2,3,\ldots.$$

What is the value of this probability for $x=0$?   (ii) Show that U is geometrically distributed with parameter $1-(1-\rho)^2$.
(iii) Show that

$$P(V-U=x)=\frac{2\rho}{2-\rho}(1-\rho)^x, \qquad x=1,2,3,\ldots.$$

What is the probability for $x=0$?
(iv) Show that U and $V-U$ are independent.

53. X and Y are geometric with parameter ρ. Let $U=Y-X$ and $V=\min(X,Y)$. (i) Show that

$$P(U=u,V=v)=\begin{cases} P(X=v-u)P(Y=v), & u<0, \\ P(X=v)P(Y=u+v), & u\geq0. \end{cases}$$

(ii) Show that
$$P(U=u,V=v)=\rho^2(1-\rho)^{2v}(1-\rho)^{|u|}.$$

(iii) Show that U and V are independent.

54. X and Y have probability generating functions $\phi(s)$, $\psi(s)$; show that $P(X-Y=x)$ is the coefficient of s^x in the expansion of $\phi(s)\psi(1/s)$ in powers of s, $x=0$, ±1, $\pm2,\ldots$.

55. Let X be geometric over the non-negative integers. (i) Show that for integers $y\le x$

$$P(X\le x\mid X\ge y)=P(X\le x-y).$$

(ii) Show that the geometric distribution is the only distribution over the non-negative integers with this property. This property is called the *memoryless* property of the geometric distribution, and can be expressed by saying that *truncating* the geometric distribution (omitting the first k probabilities and normalizing the remainder to unity) yields the same distribution. The memoryless property becomes important in Chapter 5.

56. In a certain restaurant, 90% of the customers do not smoke and 10% do. Let $X=$ number of the customer who is the fifth smoker. Write down $P(X=x)$.

57. On the student council there are four undergraduate men, six undergraduate women, and six graduate men. How many graduate women must be appointed to the council if sex and graduate status are to be independent?

58. Show how the derivation of the binomial and negative binomial distributions can be simplified by using indicator random variables.

59. A coin is weighted so that the probability of heads is one-fourth. The coin is tossed four times and $X=$ length of the longest string of tails which occurs. Find $P(X=x)$.   *Ans.* $P(X=3)=\frac{27}{128}$

60. A rodent is placed in a cage with three doors. The first door leads to food after three minutes of travel. The second door returns the creature to his starting point after 5 minutes of travel, while the third returns him to the starting point after 7 minutes of travel. What elapsed time before reaching the food would be consistent with the hypothesis that the rodent is choosing doors at random?   *Ans.* 15 minutes

61. Referring to a joint distribution function $F(x, y)$ show that

$$P(x_1\le X<x_2, y_1\le Y<y_2)=F(x_2, y_2)-F(x_1, y_2)-F(x_2, y_1)+F(x_1, y_1).$$

62. A slot machine works by inserting a coin. If a player wins, the coin is returned with another coin, otherwise the original coin is lost. The probability of winning is arranged to be one-half, independently of previous plays, unless the previous play was a win, in which case the probability of a win is $p<\frac{1}{2}$. Show that if the cost of maintaining the machine is c coins a day, then, in order to be profitable, the owner must choose p so that it satisfies the inequality

$$p<\frac{1-3c}{2(1-c)} \qquad \text{for } c<\tfrac{1}{3}.$$

63. The spores of a certain plant are arranged in sets of four in a linear chain (with three links). When the spores are ejected from the plant, each link has a probability β of breaking, independently for each link. For example, if all links break, four groups of one spore each are obtained, whereas if no links break, a single group of four spores results. Find the probability of a group containing X spores, and show that $E(X)=1+3\beta$.

64. Express $\mathrm{var}(X)$ and $\mathrm{var}(Y)$ in terms of the probability generating function $E(s^X t^Y)$.

65. For a certain species of animal, the probability of a litter of size X is Poisson with values in the positive integers, i.e.,

$$P(X=x)=\frac{e^{-\lambda}\lambda^{x-1}}{(x-1)!},\qquad x=1,2,3,\dots.$$

The probability of a male birth is p and that of a female birth, $1-p$. What is the probability that a litter is "matable," that is, that it contains at least one male and one female?

66. The number of people arriving at a library per hour is Poisson with parameter 5. An arrival is equally likely to be a man or a woman. What is the conditional probability that at most three men arrived, given that five women arrived?
    Ans. $(1/120e^5)\sum_{j=5}^{8}(\tfrac{5}{2})^j/(j-5)!$

67. Show that the correlation coefficient must lie in the interval $[-1,+1]$.

68. Verify the last formula in Section 2.7 for the geometric distribution.

69. X is Poisson with parameter λ; Y conditional on $X=x$ is binomial with parameters x and π. Show that Y is Poisson with parameter $\lambda\pi$.

70. Using the notation of Section 2.8, show that X and Y are *equidistributed* (same probabilities with the same parameters) if

$$\psi_k(s)=s^k\psi_k(1/s)\qquad\text{for all }k,$$

that is, if the conditional distribution is symmetrical,

$$q_x(n)=q_{n-x}(n).$$

Find a counterexample to show that the converse is not true.

71. Petersburg Paradox. A Bernoulli experiment with $P(S)=\tfrac{1}{2}$ is performed repeatedly. Let X be the trial on which the first success is obtained, with $P(X=x)$ being the distribution. Someone is paid 2^x when the first success occurs. Show that the expected winnings (and hence the "fair entry price") is infinite.
    [Note: This is considered a "paradox" in the sense that an infinite price is to be

paid for a finite reward. It would be more appropriate to say, however, that the "wrong" question is being asked. In a distribution with an infinite mean, a person is asked to choose a finite $E(X)$. Alternatively, one might remark that a finite value for a random variable with an infinite expectation is hardly paradoxical, since the expectation is a weighted sum.]

3

Markov Chains

3.1. Introduction: Random Walk

Markov[†] chains are the most substantial application of conditional probability which is easily accessible, and, at the same time, they provide an excellent introduction to the more general subject of stochastic processes. A stochastic process is a random variable with a time index (say, X_n, $n = 0, 1, 2, \ldots$) for discrete time, or a family of random variables (say, $X(t)$, $0 < t < \infty$) for continuous time.

The distinction between discrete and continuous is important in mathematics, as in ordinary language and in studying stochastic processes; the distinction is twofold: between time-discreteness and time-continuity on the one hand, and between discrete random variable values and continuous random variable values on the other.

The random variables in Chapters 1 and 2 are all discrete, in that the quantities involved are *counted*; in Chapter 4, the quantities will be *measured*, and thus will be continuous. The difference between counting and measuring, however clear, might seem ambiguous: Although apples are counted and water measured, one could nevertheless measure (the weight of) apples or count (the gallons of) water. Mathematically, there is no confusion; integer values are discrete, while real values are continuous.

In the time domain, it seems to be a peculiarity of the English language to use the same word "time" both for discrete time (occasion) and continuous time (duration). Thus we say that he sneezed for the fifth time or that the time he spent in bed was five hours. It is understood that the time in bed

---

[†]A. A. Markov, Russian mathematician, 1856–1922.

(duration) could be refined to minutes and seconds and is thus continuous, while the number of sneezes could not be nonintegral.

The present chapter deals with discrete variables in discrete time; the following chapter with continuous variables. Then, in Chapter 5, continuous time is introduced for both types of random variables. The word "chain" is equivalent to the assumption of discrete time; "process" usually means continuous time, although some authors use it for both. A Markov chain is one in which each random variable has a distribution which depends only on the value taken by the preceding variable and some fixed constants of the process. Thus a sequence of Bernoulli trials could define a chain of random variables, but, depending on which random variable is selected, the chain could be independent, as in Section 2.4, or Markov, as in this section.

The accumulated number of successes Y_n (a Markov chain) contrasts with X_n, the instantaneous number of successes (a Bernoulli chain), as in the example of Table 3.1.

It is easy to see that the X_n are independent—the simplest kind of chain —whereas the Y_n depend on one another only through the immediately preceding value—the next simplest kind of chain ("Markov"). The Markov property, in this example, is expressed in the fact that the probability that there are, say, six successes at any stage depends only on the number of successes accumulated at the preceding stage (it must be five or six) together with the (fixed) probability of a success or a failure on a given trial.

As an introduction to the ideas and the terminology of Markov chains, this simple example will now be expanded in a little more detail. The general formulation is given in the following sections.

Any value taken by a random variable in a chain is called the *state of the system* and the collection of all possible values, the *state space*. In the example below, the system starts in state zero ($Y_0 = 0$) before one of the trials is performed, and then gradually builds up through the positive integers to the value seven. For example, $Y_1 = 1$, $Y_2 = 2$, $Y_4 = 3$, $Y_8 = 5$, and $Y_{16} = 7$. In this chain, the state space consists of the non-negative integers.

Table 3.1. The instantaneous number of successes, X_n, vs. the accumulated number of successes, Y_n

	S	S	S	F	S	S	F	F	F	S	F	F	F	F	S	F
X_n	1	1	1	0	1	1	0	0	0	1	0	0	0	0	1	0
Y_n	1	2	3	3	4	5	5	5	5	6	6	6	6	6	7	7

From Section 2.4, it is known that

$$P(Y_n = x) = \binom{n}{x} \pi^x (1-\pi)^{n-x}, \qquad x = 0, 1, \ldots, n.$$

The experiment could be rephrased as follows: Consider a particle which starts at the origin and moves one step to the right with probability π and remains stationary with probability $1-\pi$. Each change of the system is called a *transition*; in this chain only two types of transitions are possible: from state x to state $x+1$ (with probability π) and from state x to state x (with probability $1-\pi$). Such a chain is called a *random walk* (on the positive integers); it is a particular kind of random walk, in which steps to the left or steps greater than one are impossible. In terms of conditional probability, the random walk is defined by the equations

$$P(X_{n+1} = x+1 \mid X_n = x) = \pi,$$

$$P(X_{n+1} = x \mid X_n = x) = 1 - \pi,$$

with the probability of every other kind of transition being zero.

The formulation does not include any information about where the system starts (e.g., whether $X_0 = 0$ or $X_0 = 15$), but otherwise describes completely how the chain can proceed and the probability of the various alternatives. The calculation of the (binomial) probability distribution of X_n obviously depends on the knowledge that the system started in state zero.

In addition to the ("one-step") probabilities defining the system, it is not difficult to calculate probabilities for two (or more) steps. For example,

$$P(X_{n+1} = x+2 \mid X_n = x) = \pi^2,$$

$$P(X_{n+1} = x+1 \mid X_n = x) = 2\pi(1-\pi),$$

$$P(X_{n+1} = x \mid X_n = x) = (1-\pi)^2.$$

This kind of a random walk suggests a more general one in which steps in both directions are probable, but it will be better to postpone the discussion of random walks in general until a little more mathematical technique is available.

3.2. Definitions

Most of the Markov chains considered in this chapter have a state space consisting of the non-negative integers or some subset thereof, and definitions will be given for such a state space. A sequence of random variables $X_0, X_1, X_2, \ldots$ is said to form a Markov chain if the following conditional probability equation holds for all $x, y, i_0, \ldots, i_{n-1}$, and all n:

$$P(X_{n+1}=y \mid X_n=x, X_{n-1}=i_{n-1}, \ldots, X_0=i_0)=P(X_{n+1}=y \mid X_n=x). \qquad (1)$$

This says simply that the $(n+1)$st probability distribution conditional on all preceding ones equals the $(n+1)$st probability distribution conditional on the nth for $n=0,1,2,\ldots$. Note: This definition includes the case of an independent sequence.

The definition relates each random variable in the sequence to the preceding one, and so to complete the definition of the chain it is necessary to specify a distribution for the zeroth random variable X_0. This will be denoted by $(a_0, a_1, a_2, \ldots)$, that is,

$$P(X_0=x)=a_x, \qquad x=0,1,2,\ldots.$$

When the initial state of the system is given (without any probabilistic component), $P(X_0=x)=1$ for some particular x.

The transition probabilities are often[†] abbreviated p_{xy}, where

$$p_{xy}=P(X_{n+1}=y \mid X_n=x), \qquad x, y=0, 1, 2, \ldots.$$

The Markov chain is thus completely described by the *initial probability distribution* a_x and the *transition probabilities* p_{xy}. When $X_n=x$, the chain is said to be in *state* x *at time* n; in case of a random walk it is sometimes described by saying that *the particle* is at the value x on the nth step.

Markov chains are useful in modeling a great variety of phenomena, but it is important to bear in mind that not every chain is Markov. Often the Markov property can be quite troublesome to confirm. For example, it is sometimes assumed that the weather forms a Markov chain with two states: $X_n=0$ meaning that the nth day is "dry," and $X_n=1$ meaning that the nth day is "wet," with the probability of "rain today" given "rain yesterday" being given by

[†]Note that the letters x and y occur in typographically reverse order on the two sides of the equation. Older textbooks often straightened this out by making p_{xy} the probability of a transition from y to x.

$$P(X_n = 0 | X_{n-1} = 0) = \alpha,$$

$$P(X_n = 1 | X_{n-1} = 1) = \beta.$$

This sort of modeling, while possibly useful as an intuitive introduction to the idea of a Markov chain, has the decided disadvantage that it may not correspond to meteorological reality; the probability of rain today may well depend not only on yesterday's weather, but on longer-range developments.

There are, of course, some realistic interpretations for a two-state Markov chain. The list of interpretations for "success" and "failure" given in Section 2.4 will suggest a number of possibilities. The student should think over these interpretations in connection with the Markov requirement to see which ones seem to be adequately modeled by a Markov chain, given a sequence of nonindependent success/failure symbols.

The random walks are an important category of Markov chains. A general random walk on the non-negative integers is defined by the equations

$$P(X_n = x | X_{n-1} = x - 1) = \lambda,$$

$$P(X_n = x | X_{n-1} = x) = 1 - \lambda - \mu, \tag{2}$$

$$P(X_n = x | X_{n-1} = x + 1) = \mu,$$

giving respectively the probabilities of a *step right*, *remaining in place*, and a *step left*. These transition probabilities must be completed by some statement as to what happens at the origin. It might be assumed that if the particle arrives at the origin, it stays there forever and the process terminates,

$$P(X_n = 0 | X_{n-1} = 0) = 1 \qquad (absorbing\ barrier),$$

or that it always bounces back,

$$P(X_n = 1 | X_{n-1} = 0) = 1 \qquad (reflecting\ barrier),$$

or, including both these possibilities, that

$$P(X_n = 1 | X_{n-1} = 0) = \alpha, \qquad 0 \leq \alpha \leq 1.$$

Another special random walk with some historical interest is the problem of the *gambler's ruin*. The value of X_n is the amount of the gambler's fortune after the nth game. It is assumed that he wins or loses one unit each time he gambles, so that $\lambda + \mu = 1$, and once his fortune is lost, the process terminates, $\alpha = 0$. Depending on the circumstances, the game may be *fair* ($\lambda = \mu$) or not, the gambler may stop after accumulating a predetermined fortune N (absorbing barrier at N) or he may continue indefinitely (infinitely rich adversary). As a modeling problem, it is certainly reasonable to use the Markov model, that is, to assume that his fortune after the nth game depends only on his fortune after the $(n-1)$st game and the probability of winning.[†]

Two kinds of limiting probability distributions are especially important; (i) the *equilibrium* probabilities (denoted by π_x),

$$\lim_{n \to \infty} P(X_n = x) = \pi_x,$$

which, being limiting values, may or may not exist, and (ii) the *stationary* probabilities (denoted by v_x),

$$P(X_n = x) = P(X_{n+1} = x) = v_x,$$

which also may or may not exist.

To give an example,[‡] consider the following problem: A gentleman owns three suits, green, red, and blue. He wears them on successive days according to the following scheme: If he wears the green suit one day, he is equally likely to wear it or the red suit the following day; if he wears the red suit one day, he never wears it the following day, but is equally likely to wear either of the other two; finally, if he wears the blue suit one day, he never wears the red suit the following day, but is equally likely to wear either of the other two.

With stationary probabilities v_G, v_R, v_B, Eq. (6) of Section 2.3 shows the way to the following equations:

$$\tfrac{1}{2} v_G + \tfrac{1}{2} v_R + \tfrac{1}{2} v_B = v_G,$$

$$\tfrac{1}{2} v_G \qquad\qquad = v_R,$$

$$\tfrac{1}{2} v_R + \tfrac{1}{2} v_B = v_B,$$

[†] The student may be able to think of circumstances in which this would not be a reasonable assumption.

[‡] Whimsical examples are often used in introducing Markov chains because most realistic examples involve substantial calculations. However, when the states of the system are labels rather than numbers, some fundamental tinkering with the definitions of random variable or Markov chain is needed for consistency.

which, combined with the requirement of normalization to unity, yield solutions $v_G = \frac{1}{2}$, $v_R = \frac{1}{4}$, $v_B = \frac{1}{4}$. (The need for the probabilities to sum to unity is important, as will become apparent in the algebraic treatment to follow.) These probabilities could then be interpreted as the "long-range" usage rates of the three suits. The student should note two important facts. First, the calculations of the stationary probabilities have been accomplished without any assumptions regarding the initial vector—the probability of red, green, or blue on the first day of the experiment. Second, although the calculations have been of the stationary probabilities, the interpretation of the result is as if the equilibrium probabilities were found. Thus it seems plausible that a set of stationary probabilities could be equilibrium, and this fact will be established later in the chapter. When the theorem is proven with appropriate restrictions (in Section 3.8), the plausible will become true.

Given the transition matrix and the initial vector, it is in principle possible to calculate any quantity required. For example, the distribution of X_1 can be written

$$P(X_1 = x) = \sum_{j=0}^{\infty} P(X_0 = j)P(X_1 = x \mid X_0 = j)$$

$$= \sum_{j=0}^{\infty} a_j p_{jx}.$$

But the calculations leading to distributions for the other random variables in the chain $P(X_n = x)$ can often become difficult algebraically.

To examine systematically the various distributions implicit in the definition of a Markov chain, it is highly desirable to use the notation (and properties) of matrix and vector.

3.3. Matrix and Vector

The student familiar with matrix theory will recognize in the calculations of Section 3.2 the elements of matrix transformation of vectors. The formulation in terms of matrices makes the general results more compact and assists in the proof of theorems. In the present section, a brief review outline of the necessary portions of the theory is given, together with a theorem important in Markov chain analysis. The matrices are assumed to be square, but not necessarily finite.

A (square) *matrix* is an array of numbers arranged in rows and columns; if the number of rows and columns is finite, then they are equal; if one is infinite, then so is the other. The individual numbers are called *elements* of the matrix. A matrix with all row sums equal to unity and all elements positive or zero is called a *stochastic matrix*. Two matrices are *equal* only if corresponding elements are equal. Addition of matrices is defined by adding corresponding elements and only when they have the same number of rows and columns. Multiplication of matrices is defined in the same circumstances by the formula

$$\mathbf{AB} = \{a_{xy}\}\{b_{xy}\} = \left\{ \sum_j a_{xj} b_{jy} \right\},$$

where a_{xy} denotes the element in the xth row and yth column of $\mathbf{A}$, and $\{a_{xy}\}$ is the matrix of which a_{xy} is a typical element.

Vectors are of two types, row vectors and column vectors; vectors are not necessarily finite. A *row vector* is an ordered sequence of numbers written in a row; a *column vector* is an ordered sequence of numbers written in a column. The numbers are called *components* of the vector. Vectors are equal only if both are of the same type and have equal components. A *probability vector* has all components zero or positive, with the sum of the components equal to unity. (Thus the rows of a stochastic matrix are probability vectors.) Vectors are added by adding corresponding components and then only if both are of the same type, with the same number of components.

Vectors are multiplied by matrices only if the number of components of the vector equals the number of rows (column) of the matrix, and then by means of the formula

$$\mathbf{uA} = \left(\Sigma u_j a_{j1}, \Sigma u_j a_{j2}, \Sigma u_j a_{j3}, \ldots \right),$$

where $\mathbf{u}$ is a row vector, and

$$\mathbf{Av} = \begin{pmatrix} \Sigma v_j a_{1j} \\ \Sigma v_j a_{2j} \\ \vdots \end{pmatrix},$$

where $\mathbf{v}$ is a column vector.

Table 3.2. Results of various kinds of multiplication

	Scalar	Row vector	Column vector	Matrix
Scalar	Scalar	Row vector	Column vector	Matrix
Row vector	Row vector	Undefined	Scalar	Row vector
Column vector	Column vector	Matrix	Undefined	Undefined
Matrix	Matrix	Undefined	Column vector	Matrix

If, in vector–matrix multiplication, **u** is a row vector with zero in every position except for the jth, and $+1$ in the jth position, then **uA** is the jth row of **A**. Similarly, if **v** is a column vector with zero in every position except the jth and $+1$ in the jth position, then **Av** is the jth column of **A**. In these cases the vectors are called the (jth) *unit vectors*.

The products of a row vector **u** and a column vector **v** are defined only when the two vectors have the same number of components. Then the product **uv** is the number $\Sigma u_j v_j$ and the product **vu** is the matrix $\{v_x u_y\}$.

In the context of vectors and matrices, a number is called a *scalar*. Multiplication of vectors and matrices by scalars is defined by multiplying each component or element by the scalar. The results of various kinds of multiplication are set out in Table 3.2.

Vectors **u** and **v** satisfy $\mathbf{u} \le \mathbf{v}$ if and only if they have the same number of components, and corresponding components satisfy $u_x \le v_x$.

Theorem 1. Let $\mathbf{A} = \{a_{xy}\}$ be a finite $(r \times r)$ stochastic matrix having no zero elements, and let β be the smallest element $(\ne 0)$. Let v be any r-component column vector, with largest component M_0 and smallest component m_0. Let M_1 and m_1 be respectively the largest and smallest components of **Av**. Then

(i) $\qquad M_1 \le M_0,$

(ii) $\qquad m_1 \ge m_0,$

(iii) $\qquad M_1 - m_1 \le (1 - 2\beta)(M_0 - m_0).$

Proof. Define an r-component column vector $\mathbf{w}$ by replacing every element of $\mathbf{v}$, except for one of the m_0 elements, by M_0. Then $\mathbf{w} \geq \mathbf{v}$. Multiply $\mathbf{A}$ by $\mathbf{w}$, to obtain

$$
\begin{pmatrix} a_{11} & \cdots & a_{1r} \\ \vdots & & \vdots \\ a_{r1} & \cdots & a_{rr} \end{pmatrix} \begin{pmatrix} M_0 \\ \vdots \\ m_0 \\ \vdots \\ M_0 \end{pmatrix} = \begin{pmatrix} M_0(1-a_{1k})+m_0 a_{1k} \\ M_0(1-a_{2k})+m_0 a_{2k} \\ \vdots \\ M_0(1-a_{rk})+m_0 a_{rk} \end{pmatrix},
$$

where k is the number of the element of $\mathbf{w}$ having the value m_0. Thus every element of $\mathbf{A}\mathbf{w}$ is of the form

$$
M_0 - a(M_0 - m_0),
$$

where of course the value of a is different for each component, but always with $a \geq \beta$. Thus every component of $\mathbf{A}\mathbf{w}$ is $\leq M_0 - \beta(M_0 - m_0)$, and since no component of $\mathbf{A}\mathbf{v}$ can exceed a component of $\mathbf{A}\mathbf{w}$, the largest such component, M_1 must satisfy

$$
M_1 \leq M_0 - \beta(M_0 - m_0),
$$

and this inequality is stronger than (i). Applying the same argument to the vector $-\mathbf{v}$ gives the corresponding inequality

$$
-m_1 \leq -m_0 - \beta(-m_0 + M_0),
$$

which establishes (ii). The result (iii) is obtained by adding the two inequalities above.   □

Note that the proof breaks down in the case of an infinite matrix simply because it could happen that no least element would exist. It is left as an exercise for the student to discover why the condition is imposed that the matrix have no zeros.

It is also interesting to note that the calculation of the probabilities $P(X_1 = x)$ given in Section 3.2 is exactly equivalent to the multiplication of the transition matrix by the initial row vector.

3.4. The Transition Matrix and Initial Vector

The definitions of Section 3.2 show that a Markov chain is specified by a state space, a collection of transition probabilities together with an initial distribution:

$$p_{xy} = P(X_{n+1} = y | X_n = x),$$

$$a_x = P(X_0 = x).$$

These quantities can be considered a (finite or infinite, but square) matrix

$$\mathbf{P} = \{p_{xy}\}$$

and a (finite or infinite) row vector

$$\mathbf{a} = (a_0, a_1, \ldots).$$

Whether $\mathbf{P}$ and $\mathbf{a}$ are finite or infinite depends on the number of states possible for the chain, that is, the number of values possible for the random variables X_n. It is convenient to consider the number to be the same for all n; if some are in fact impossible, the assignment of probability zero takes care of the situation.

Example 1 (Random Walk with Absorbing Barrier).

$$\mathbf{P} = \begin{pmatrix}
1 & 0 & 0 & 0 & 0 & \cdots \\
\mu & 1-\lambda-\mu & \lambda & 0 & 0 & \cdots \\
0 & \mu & 1-\lambda-\mu & \lambda & 0 & \cdots \\
0 & 0 & \mu & 1-\lambda-\mu & \lambda & \cdots \\
& \vdots & & & \vdots & \cdots
\end{pmatrix}.$$

Example 2 (Random Walk with a Reflecting Barrier). The transition matrix $\mathbf{P}$ would be the same as in Example 1 except that the first row would be $(0, 1, 0, 0, 0, \ldots)$.

Example 3 (Green/Red/Blue Example from Section 3.2). Assume that the rows and columns are respectively the colors in the order given above:

$$\mathbf{P} = \begin{pmatrix} \frac{1}{2} & \frac{1}{2} & 0 \\ \frac{1}{2} & 0 & \frac{1}{2} \\ \frac{1}{2} & 0 & \frac{1}{2} \end{pmatrix}.$$

Example 4 (Fish-Catching Example from Section 2.1, with the state of the system defined to be the number of fish caught on the first n scoops).

$$\mathbf{P} = \begin{pmatrix} (1-\pi)^3 & 3\pi(1-\pi)^2 & 3\pi^2(1-\pi) & \pi^3 \\ 0 & (1-\pi)^2 & 2\pi(1-\pi) & \pi^2 \\ 0 & 0 & 1-\pi & \pi \\ 0 & 0 & 0 & 1 \end{pmatrix}.$$

In these examples, the initial vector would be determined by the nature of the hypothesis. For Example 1, the random walker could be assumed to start n units away from the origin:

$$\mathbf{a} = (0,0,0,\ldots,1,\ldots,0,0,\ldots).$$

In Example 3, the person might be assumed to begin his bizarre costuming procedure by selecting a suit at random,

$$\mathbf{a} = \left(\tfrac{1}{3}, \tfrac{1}{3}, \tfrac{1}{3}\right),$$

while in Example 4, the initial number of fish caught is given to be zero:

$$\mathbf{a} = (1,0,0,0).$$

With the formulation as transition matrix and initial vector, the calculation of probabilities for the random variable X_1 can be seen as vector-by-matrix multiplication:

$$P(X_1 = x) = \Sigma a_j p_{jx},$$

that is, $\mathbf{aP}$ is a row vector with xth component $P(X_1 = x)$. Thus, also, the stationary vector $\mathbf{v}$ satisfies the equation

$$\mathbf{vP} = \mathbf{v}.$$

These equations are written out in full in Section 3.2 for the colored-suit example, and the student should confirm that they correspond to the matrix given above.

In general, the stationary probabilities, although simply represented, are easy or difficult to compute, depending on whether or not the equations $\mathbf{vP}=\mathbf{v}$ are easy or difficult to solve. The equilibrium probabilities, on the other hand, present further difficulties, which will be discussed in the next section.

The examples given in this section slide over the question of state space, mainly because the non-negative integers are assumed to correspond naturally to the rows and columns of the transition matrix. Where some other state space is intended (as in the colored-suit example), it is, strictly speaking, necessary to label the rows with the states to which they are supposed to correspond. Thus a given stochastic matrix could apply to two different Markov chains, just as in Chapter 1 the same set of probabilities could apply to different random variables. In Example 3, the precise way of writing the matrix should be

$$
\begin{array}{c}
G \\
R \\
B
\end{array}
\left(
\begin{array}{ccc}
\frac{1}{2} & \frac{1}{2} & 0 \\
\frac{1}{2} & 0 & \frac{1}{2} \\
\frac{1}{2} & 0 & \frac{1}{2}
\end{array}
\right),
$$

because there could be defined another Markov chain with state space 1, 2, 3 and transition matrix

$$
\begin{array}{c}
1 \\
2 \\
3
\end{array}
\left(
\begin{array}{ccc}
\frac{1}{2} & \frac{1}{2} & 0 \\
\frac{1}{2} & 0 & \frac{1}{2} \\
\frac{1}{2} & 0 & \frac{1}{2}
\end{array}
\right).
$$

Inasmuch as this chapter deals mostly with non-negative integral state spaces, the distinction is often neglected.

3.5. The Higher-Order Transition Matrix: Regularity

The probability of a transition in two steps from state x to state y of a Markov chain can be calculated by conditioning on the state of the chain at the intermediate step. Using Eq. (9) of Chapter 2,

$$P(X_{n+2}=y \mid X_n=x) = \sum_j P(X_{n+2}=y \mid X_{n+1}=j) P(X_{n+1}=j \mid X_n=x)$$

$$= \sum_j p_{xj} p_{jy}, \tag{3}$$

which is exactly the formula for matrix multiplication. Let this probability be denoted by $p_{xy}^{(2)}$. Then the matrix $\{p_{xy}^{(2)}\}$ is just the square of the matrix $\{p_{xy}\}$. This can be extended by the same argument to higher-order transition probabilities. If $p_{xy}^{(n)}$ denotes the probability of a transition from state x to state y in exactly n steps, then $\{p_{xy}^{(n)}\} = \mathbf{P}^n$. Thus the matrix multiplication formula

$$\mathbf{P}^m \mathbf{P}^n = \mathbf{P}^n \mathbf{P}^m = \mathbf{P}^{m+n} \tag{4}$$

can be interpreted for the Markov chain transition matrix as stating that the probability of a transition from state x to state y (where x and y are arbitrary) in $m+n$ steps is the same as the probability of an m-step transition to some intermediate state followed by an n-step transition to y. In this context, the matrix multiplication equation is called the *Chapman–Kolmogorov equation for the Markov chain*. This equation will be important in the sequel, and it is important to realize that the proof of the Chapman–Kolmogorov equation consists of nothing more than the identification of the matrix multiplication formula with the probabilistic description of the Markov chain.

If there is an integer n such that $\mathbf{P}^n$ consists entirely of positive (nonzero) elements, then the matrix and the Markov chain it represents are called *regular*. In a Markov chain with a regular transition matrix, every state must be *accessible* from every other state, and in the same number of steps. It would be impossible in a regular Markov chain that there could exist two states x and y with $p_{xy}^{(n)} = 0$ for all n. Also, regular chains could not have states x and y accessible only in an even number of steps, because if $\mathbf{P}^n$ consists only of positive elements, so will higher powers of $\mathbf{P}$. Finally, regularity is incompatible with an infinite matrix such as the one characterizing the random walk, since it is impossible that every transition can occur in a finite number of steps, and n must be a finite integer.

Thus regularity is a rather stringent condition on the Markov chain, and when it is satisfied, produces quite simple general results. The following theorems state in essence that for finite regular chains, the equilibrium probabilities exist, and are equal to the stationary probabilities.

Theorem 1. For a finite regular transition matrix $\mathbf{P}$, $\lim \mathbf{P}^n$ exists and every row of the limiting matrix is identical.

Proof. Case I: No zeros in $\mathbf{P}$. Let M_n and m_n be the maximum and minimum components of the column vector $\mathbf{P}^n \mathbf{U}_j$, where $\mathbf{U}_j$ is the jth unit (column) vector. Then, by the theorem of Section 3.3, $M_n \geq M_{n+1}$ and $m_n \leq m_{n+1}$ for $n = 1, 2, 3, \ldots$. Also, by the same theorem,

$$\Delta_n = M_n - m_n \leq (1 - 2\beta)\Delta_{n-1}, \qquad n = 1, 2, 3, \ldots,$$

where β is a nonzero minimum element of $\mathbf{P}$. In the present case, $\mathbf{P}$ is a stochastic matrix, and so $\beta \leq \frac{1}{2}$. The inequalities imply that

$$\Delta_n \leq (1 - 2\beta)^n \Delta_0 \leq (1 - 2\beta)^n.$$

As $n \to \infty$, $\Delta_n \to 0$, and $\mathbf{P}^n \mathbf{U}_j$, the jth column of $\mathbf{P}^n$, approaches a column vector with all components equal.

Case II: Zeros in $\mathbf{P}$. Let N be such that $\mathbf{P}^N$ has no zeros, and let β_N be the smallest (nonzero) element of $\mathbf{P}^N$. For matrices $\mathbf{P}^{kN}$ which are multiples of $\mathbf{P}^N$, case I of the theorem holds, and therefore the nonincreasing sequence Δ_n contains a subsequence Δ_{kN} which approaches zero. Thus Δ_n approaches zero and the theorem is proven. $\square$

Theorem 2. If $\mathbf{P}$ is a finite regular transition matrix with $\lim \mathbf{P}^n = \Pi$, where each row of Π is $\boldsymbol{\pi} = (\pi_x)$, then for any initial probability (row) vector $\mathbf{a} = (a_x)$, $\mathbf{a}\mathbf{P}^n \to \boldsymbol{\pi}$, and $\boldsymbol{\pi}$ is the unique stationary vector of $\mathbf{P}$.

Proof. Direct multiplication shows that $\mathbf{a}\Pi = \boldsymbol{\pi}$; hence $\mathbf{a}\mathbf{P}^n \to \boldsymbol{\pi}$. Furthermore, if $\boldsymbol{\sigma}$ is a stationary vector (such that $\boldsymbol{\sigma}\mathbf{P} = \boldsymbol{\sigma}$), then, since $\boldsymbol{\sigma}\mathbf{P}^n \to \boldsymbol{\pi}$ and $\boldsymbol{\sigma}\mathbf{P}^n = \boldsymbol{\sigma}$, $\boldsymbol{\sigma} = \boldsymbol{\pi}$ so that $\boldsymbol{\pi}$ is the unique stationary vector of the Markov chain. The proof is completed. $\square$

The proofs given above apply only to finite regular chains, but the theorems remain true if the condition of finiteness is omitted. Since the proofs shed little light on probabilistic problems, they are omitted.

These results show that the long-range behavior of regular chains can be determined relatively easily by calculation of the stationary vector. The calculation itself may present difficulties, but these difficulties are algebraic rather than probabilistic, and are not discussed in this volume.

When a Markov chain is not regular, the situation is somewhat more complicated and can most easily be described after some preliminary definitions.

3.6. Reducible Chains

In classifying Markov chains, a first step is to remove from consideration certain types of chains. Consider a game in which children throw a ball from one to another, and let the chain be in state x when the xth child has the ball. (This is a rather common image for a chain with zeros on the diagonal, assuming that no child throws the ball to itself.) Given the Markov hypothesis and the probabilistic behavior of the children, it would be easy to write down the transition matrix and perhaps calculate the stationary probabilities, which would then be interpreted as the probabilities that each of the children had the ball. However, suppose further that girls throw only to girls and boys to boys. Then the development of the game depends on who gets the ball first, and depending on this outcome, either of two separate chains develops. Such a chain is called reducible; the exact definition will follow. The simplest reducible chain has the matrix

$$\begin{pmatrix} 1 & 0 \\ 0 & 1 \end{pmatrix},$$

which leads to the equations for stationary probabilities

$$v_0 = v_0,$$

$$v_1 = v_1,$$

from which it is not possible to infer a unique solution, even by adding the normalizing condition $v_0 + v_1 = 1$.

In general, a reducible chain has a matrix $\mathbf{P}$ which can be decomposed into two submatrices $\mathbf{P}_1$ and $\mathbf{P}_2$ thus:

$$\mathbf{P} = \begin{pmatrix} \mathbf{P}_1 & 0 \\ 0 & \mathbf{P}_2 \end{pmatrix}.$$

The number of separate blocks can be greater than two. It is clear just by looking at this matrix that the Markov chain really consists of two Markov chains which do not relate to one another. Therefore it is sensible to consider each separately, with some probabilistic scheme to determine which chain is operative.

Sometimes, by renumbering the states, a reducible chain is a little more difficult to spot. Consider the matrix

$$\begin{pmatrix} 0 & \frac{1}{2} & 0 & \frac{1}{2} & 0 & 0 \\ \frac{1}{2} & 0 & 0 & \frac{1}{2} & 0 & 0 \\ 0 & 0 & 0 & 0 & \frac{1}{2} & \frac{1}{2} \\ \frac{1}{2} & \frac{1}{2} & 0 & 0 & 0 & 0 \\ 0 & 0 & \frac{1}{2} & 0 & 0 & \frac{1}{2} \\ 0 & 0 & \frac{1}{2} & 0 & \frac{1}{2} & 0 \end{pmatrix}.$$

By tracing the probabilities, the student will confirm that this corresponds to the ball-throwing story above, with three boys and three girls.

Reducible Markov chains can be exactly defined by introducing the concept of a *closed set of states*.

Definition 1. A collection C of states is called a closed set of states if

$$P(X_n \in C \mid X_{n-1} = x) = 1$$

for all $x \in C$.

Definition 2. A Markov chain is called *irreducible* if it does not contain more than one closed set of states.

Definition 3. If a closed set of states consists of a single state, this state is called an *absorbing* state.

Definition 4. A Markov chain which is not irreducible is called reducible.

From this point onwards, it is assumed that all chains are irreducible.

3.7. Periodic Chains

A second kind of Markov chain for which it is desirable to treat separately, and often to exclude from the discussion, is the periodic chain, that is, a chain with periodic states.

Definition 1. A state x is called *periodic* if $p_{xx}^{(n)} = 0$, except when n is a multiple of an integer greater than one.

The simplest example of a periodic chain is one which oscillates between states, with matrix

$$\begin{pmatrix} 0 & 1 \\ 1 & 0 \end{pmatrix}.$$

Obviously this chain has stationary vector $(\frac{1}{2}, \frac{1}{2})$, giving an example of a Markov chain which is not regular, and yet which has a stationary vector. Many periodic chains have this property. One of the most important such chains is discussed in detail in Section 3.11.

Most of the results in the remainder of this chapter are given for aperiodic (that is, not periodic) chains, because the main principles can be illustrated on aperiodic chains.

3.8. Classification of States. Ergodic Chains

Given a state x, let ξ be the probability that, beginning in state x, the system ever returns to the state x, $0 \le \xi \le 1$. If $\xi < 1$, the state is called *transient*, and if $\xi = 1$, the state is called *recurrent*.

Transient states are visited a finite number of times, since at each visit there is a nonzero probability $1 - \xi$ of never returning. Let the number of visits to a transient state be the random variable X. Then X is geometrically distributed with parameter ξ:

$$P(X=x) = (1-\xi)\xi^x, \qquad x = 0, 1, 2, \dots.$$

The proof of this fact consists essentially in observing that each return to the state is an independent event.

Recurrent states are further classified according to whether their *mean recurrence time* is finite or infinite. Probability distributions with infinite means were briefly introduced in Section 1.6; this is the first serious encounter with such distributions. Suppose a chain starts in state x. Let Y denote the number of steps before the chain is again (for the first time) in state x. Then Y is a random variable defined over the positive integers. Let

$$f_x(y) = P(Y = y \text{ for state } x)$$

and

$$\mu(x) = \sum_{j=1}^{\infty} j f_x(j).$$

Then $\mu(x)$ is called the *mean recurrence time* for state x.

A recurrent state x is called *ergodic* if $\mu(x) < \infty$, that is, if the series defining $\mu(x)$ converges, and *null* if the series diverges.[†]

Theorem 1. If two states x and y are accessible from one another, then they are either both transient, both ergodic, or both null.

Proof. Case I: One State Transient. Suppose x is transient and let y be recurrent. Let p be the probability of passing through state x going from state y back to state y. Since x is accessible from y, $p > 0$. The process returns to state y infinitely often, and so the transition from y to x is available infinitely often and, having a nonzero probability, must take place infinitely often. Thus x is recurrent, which is a contradiction. Thus y must also be transient.

Case II: One State Ergodic. Suppose y is ergodic, and p is defined as in case I. Then, since the mean number of steps away from y is finite and each visit carries a nonzero probability of a visit to x also, the mean time away from x on those loops which pass through y must also be finite. Even if the mean recurrence times on other loops were infinite, the mean recurrence time $\mu(x)$ would still be finite.

Case III: One State Null. This is established by consideration of the other two cases: If x is null, then y could be neither ergodic nor transient. □

In the process of cumulative Bernoulli counts, all states are transient, since there is a finite probability that any state will be left and once left, never revisited. In the fish-netting process, with the random variable being

---

[†] The terms "ergodic" and "null," although not particularly intuitive, are standard.

the cumulative number of fish caught, states 0, 1, and 2 are transient, but state 3 is ergodic and, in fact, absorbing.

Note that the theorem does not prevent a chain from containing more than one kind of state, but it does prevent that for states which are *mutually accessible*. For an irreducible chain, there can be at most one set of mutually accessible states, although there may be some transient states as well.

Theorem 2. A finite Markov chain with all states ergodic is regular, and so the conclusions of Section 3.5 follow. (Such a chain is called an *ergodic chain*.)

Proof. Since a state x is ergodic, there must be some power N of the transition matrix $\mathbf{P}$ so that the diagonal element $p_{xx}^{(N)} \neq 0$. By the rules of matrix multiplication, since this nonzero element was produced with (possibly) zeros in lower powers, all higher powers will also have a nonzero element in this position: $p_{xx}^{(n)} > 0$ for $n = N, N+1, N+2, \dots$. For any other state y, there must be a power N' of the matrix such that $p_{xy}^{(N')} > 0$, since y, as well as x, is ergodic. Thus, since there is a positive probability of going from x to x in N steps and from x to y in N' steps, there is a positive probability of going from x to y in $N+N'$ steps, and for the power $N+N'$, both elements $p_{xx}^{(N+N')}$ and $p_{xy}^{(N+N')}$ are positive. Again, by the rules of matrix multiplication, all subsequent powers of $\mathbf{P}$ will contain positive entries in these positions. If this argument is applied successively to the (finite) collection of states, it can be established that there is a power of the transition matrix for which all elements are positive, proving regularity. $\square$

Thus, for ergodic Markov chains, the key to understanding the long-range behavior of the chain lies in calculating the stationary probabilities of the transition matrix. To summarize, for a finite irreducible chain with all states ergodic, the limits $\pi_y = \lim_{n \to \infty} p_{xy}^{(n)}$ exist, are independent of x, and satisfy the equations $\pi_x = \sum_j \pi_j p_{jx}$.

Finally, it is intuitively clear that π_x is the reciprocal of $\mu(x)$, but the proof is difficult and involves a basic theorem about infinite series. Some further discussion of the point will be given in Section 3.16 and a method will be discovered for finding the first-passage probabilities $f_x(y)$.

3.9. Finding Equilibrium Distributions—The Random Walk Revisited

When a Markov chain has only a finite number of states, the general theory applies and the problem of determining stationary vectors is the familiar algebraic problem of finding solutions to a linear homogeneous set of equations, with the additional condition of normalization to unity. That condition alone rules out both the trivial solution ($\pi_0 = \pi_1 = \pi_2 = \cdots = \pi_n = 0$) and also proportional families of solutions ($\pi_x = k\sigma_x$ for all x).

In the case of an infinite number of states, the theory is much the same, but the proofs become more complicated. In this section and the next, some examples are given of how equilibrium probabilities can be calculated if it is assumed that these probabilities exist. First, the general problem of the random walk is discussed; consider the transition matrix of Section 3.4, but with first row $(1-\alpha, \alpha, 0, 0, \ldots)$. The equilibrium probabilities (π_0, π_1, $\pi_2, \ldots$) must satisfy the infinite set of equations

$$\begin{aligned}
(1-\alpha)\pi_0 + \mu\pi_1 &= \pi_0, \\
\alpha\pi_0 + (1-\lambda-\mu)\pi_1 + \mu\pi_2 &= \pi_1, \\
\lambda\pi_1 + (1-\lambda-\mu)\pi_2 + \mu\pi_3 &= \pi_2, \\
\lambda\pi_2 + (1-\lambda-\mu)\pi_3 + \mu\pi_4 &= \pi_3, \\
\vdots \qquad\qquad \vdots
\end{aligned} \tag{5}$$

The equations could be solved recursively, and the general formula obtained by mathematical induction. However, it is often convenient to use probability generating functions in problems of this sort, and since this is the first encounter with infinite sets of equations, the method will be used, with some attention to the details. The first equation is multiplied by s^0, the second by s^1, the third by s^2, and so forth, and the resulting set of equations summed. Clearly the right side becomes simply

$$\phi(s) = \sum_j s^j \pi_j,$$

the probability generating function of the equilibrium distribution. The terms on the left side are less tidy, but can still be simplified with a little effort. The first two equations are irregular, and their contribution to the sum is written down unchanged:

$$(1-\alpha)\pi_0 + \mu\pi_1 + s\alpha\pi_0 + s(1-\lambda-\mu)\pi_1 + \mu s\pi_2.$$

The remaining equations give three kinds of terms, the first being

$$\lambda\left(s^2\pi_1 + s^3\pi_2 + s^4\pi_3 + \cdots\right),$$

which can be written

$$s\lambda[\phi(s)-\pi_0].$$

The second group of terms is

$$(1-\lambda-\mu)(s^2\pi_2+s^3\pi_3+\cdots),$$

which is

$$(1-\lambda-\mu)[\phi(s)-\pi_0-s\pi_1].$$

Finally, the third series of terms is

$$\mu(s^2\pi_3+s^3\pi_4+\cdots),$$

which can be written

$$\mu\frac{1}{s}[\phi(s)-\pi_0-s\pi_1-s^2\pi_2].$$

If the four contributions to the equation are assembled, the terms involving π_1 and π_2 cancel, and $\phi(s)$ can be expressed in terms of π_0 (and the parameters) as follows:

$$\phi(s)=\frac{-\alpha+\alpha s-\lambda s+\lambda+\mu-\mu/s}{-\lambda s+\lambda+\mu-\mu/s}\pi_0. \tag{6}$$

At this stage it is often wise to confirm that $\phi(0)=\pi_0$ and $\phi(1)=1$, as a precaution against possible error. In the present case, both of these conditions yield indeterminant forms. The first one quickly reduces to the required value, and the second indeterminant form, evaluated with L'Hôpital's rule, gives a value for π_0:

$$\pi_0=\frac{\mu-\lambda}{\mu-\lambda+\alpha}. \tag{7}$$

Equation (7), substituted back into the expression for $\phi(s)$, yields the probability generating function explicitly in terms of the parameters of the model. A careful inspection of the fractional representation of $\phi(s)$ reveals a common factor $(1-1/s)$ in the numerator and denominator. When $s\neq1$, this can be removed, giving the final form

$$\phi(s) = \frac{\mu - \lambda}{\mu - \lambda + \alpha} \frac{(\lambda - \alpha)s - \mu}{\lambda s - \mu}, \tag{8}$$

and in this form $\phi(0) = \pi_0$, $\phi(1) = 1$ are easy to confirm.

This has been a lengthy calculation, going from the transition matrix to the probability generating function of the equilibrium distribution, but the result contains a great deal of information about the Markov chain. In interpreting the result, keep in mind that

$$\lambda = P(\text{a step to the right, except from the origin}),$$

$$1 - \lambda - \mu = P(\text{standing still, except at the origin}),$$

$$\mu = P(\text{a step to the left, except at the origin}),$$

$$1 - \alpha = P(\text{standing still at the origin}),$$

$$\alpha = P(\text{a step to the right from the origin}).$$

The case $\alpha = 0$ can be disposed of first. Then the origin is an absorbing state, with $\pi_0 = 1$. It is also clear that $\pi_0 = 1$ can never occur unless $\alpha = 0$. Suppose therefore that $\alpha > 0$. The probability π_0 will be positive only if both the numerator and denominator are positive, or both negative. It is easy to see that the latter is impossible, for, since $\pi_0 < 1$, multiplying by the (negative) denominator gives

$$\mu - \lambda > \mu - \lambda + \alpha$$

or $\alpha < 0$.

On the other hand, if both numerator and denominator are positive, then the first inequality includes the second one and $\mu > \lambda$, so that there is a greater probability of a step to the left than of a step to the right.

The second kind of information available from a knowledge of $\phi(s)$ is the values of the moments. Differentiating with respect to s and setting $s = 1$ gives

$$m = \phi'(1) = \frac{\alpha \mu}{(\mu - \lambda)(\mu - \lambda + \alpha)}.$$

Similar calculations give $\phi''(s)$ and hence the variance.

Furthermore, the exact probabilities can be obtained from the probability generating function by the formula $\pi_x = \phi^{(x)}(0)/x!$ [Eq. (21) in Section 1.11]. In looking for a power-series expansion of $\phi(s)$, three factors must be considered: the constant $(=\pi_0)$, the linear term in the numerator, and the linear term in the denominator. The latter can be expanded as a geometric series:

$$\frac{1}{1-\rho s} = 1 + \rho s + \rho^2 s^2 + \rho^3 s^3 + \cdots,$$

where $\rho s < 1$. To put the denominator into this form, divide by $-\mu$ so that

$$\rho = \lambda/\mu.$$

Division by $-\mu$ makes the numerator of the form $As + B$, where

$$A = \frac{\alpha - \lambda}{\mu} \text{ and } B = 1.$$

Thus π_x, the coefficient of s^x, consists of two terms, the first being $\pi_0 A \rho^{x-1}$ and the second being $\pi_0 B \rho^x$. Making the necessary substitutions gives

$$\pi_x = \frac{\mu - \lambda}{\mu - \lambda + \alpha} \frac{\alpha}{\mu} \rho^{x-1}.$$

This formula does not apply to π_1; the student should derive that expression separately, finding *en route* why the calculation for the general case breaks down.

The general random walk results can be used to obtain a number of interesting and useful special cases, simply by assigning specific values to the parameters. Two examples will be given, the first being a (discrete time) queue, or waiting line, where the random variable represents the number in the system (waiting or being served). Queues provide important models of various kinds of stochastic processes and will occur with increasing frequency (and increasing complexity) in the remainder of this volume. The present model is a rather primitive one, because of the discrete time aspect. Customers are supposed to arrive only at discrete intervals and to be finished with service also at the same discrete intervals. The student might think of a service facility which can admit new customers only "on the hour" and can discharge customers also only "on the hour." The number of customers in the system can be $0, 1, 2, \ldots$, and the parameters are interpreted as follows: λ=probability of a new customer in the system, μ= probability of a service finishing, $\alpha = \lambda$, with π_x being the long-range frequencies of the number of customers in the queue. With this interpretation, for example, $\pi_0 = 1 - \rho$ gives the probability of an idle server.

The second interpretation, gambling with a fixed stake, was already mentioned in Section 3.2 and will be further enlarged in Section 3.14. With a merciless opponent and a probability of winning no greater than that of losing, being bankrupt is clearly an absorbing state. For a less trivial result, suppose that a generous opponent gives a bankrupt gambler a unit stake, so that the origin is a reflecting barrier, but suppose that the odds are against the gambler, so that λ, the probability of winning, is less than one-half. The student can verify that the gambler's average fortune is

$$\frac{1}{2(1-2\lambda)}.$$

It is also left as an exercise to show that for both the queueing model and the gambling model, the equilibrium distribution, if it exists, is of a modified geometric form, that is, the probabilities form a geometric series after the first two terms.

Finally, it will be a useful exercise for the student to classify the states of the random walk and to see how the classification (transient/ergodic/null) depends on the values of the parameters λ, μ, and α.

The stages of the calculation, represented by Eqs. (5), (6), (7), and (8), are quite typical of calculations which will recur in the remainder of this book, and it will be helpful for further understanding for the student to rehearse the various steps until they begin to become intuitively clear.

3.10. A Queueing Model

A random walk can be interpreted as a queue only with some difficulty, but there is a Markov chain which provides a reasonably good model for the number present in a simple queue. This chain is obtained by making the discrete times those moments when a service is completed and assuming the existence of a probability distribution

$$a_x = P(X=x) = P(x \text{ arrivals during a service period}).$$

The (naturally continuous) time for a queue is thus "discretized" by taking service periods. The model is clearly not a random walk, since a transition is possible from state x to any state $x+y$ if there are $y+1$ (allowing for the one person leaving after service) arrivals during a service period.

The student will greatly improve his understanding of this model and prepare for the more difficult models of Chapter 6 by verifying the transition matrix for the system:

$$
\begin{pmatrix}
a_0 & a_1 & a_2 & a_3 & a_4 & \cdots \\
a_0 & a_1 & a_2 & a_3 & a_4 & \cdots \\
0 & a_0 & a_1 & a_2 & a_3 & \cdots \\
0 & 0 & a_0 & a_1 & a_2 & \cdots \\
0 & 0 & 0 & a_0 & a_1 & \cdots \\
\vdots & \vdots & \vdots & \vdots & \vdots &
\end{pmatrix} . \tag{9}
$$

It will also be a good exercise to confirm that for no probability distribution a_x is this a random walk and to understand the difference in terms of the basic model.

In classifying the states of the queueing chain, let ζ denote (as in Section 3.8) the probability that, beginning in state zero, i.e., with no customers in the system, the queue will ever again return to state zero. Suppose that in the first time unit during which arrivals occur, the system jumps to state y, with $y=1,2,3,\ldots$, the probability of which is a_y. Let ζ_y be the probability that, beginning in state y, the system is ever again empty. Then returning to state zero (from state zero) can happen either by no arrivals in the first time period (probability a_0) or by jumping to state y and gradually returning to state zero, that is,

$$
\zeta = a_0 + \sum_{y=1}^{\infty} a_y \zeta_y .
$$

Since customers are served one at a time, the queue must pass through all states between y and 0 if it is to return to zero. Furthermore, the arrival and service pattern that would reduce the queue eventually from state y to state $y-1$ is exactly the same as the one that would reduce the queue from state $y-1$ to state $y-2$, and so forth (for example, seven arrivals in one time unit followed by eight time units with no arrivals). Thus

$$
\zeta = a_0 + \sum_{y=1}^{\infty} a_y \zeta_1^y .
$$

Now, ζ_1, the probability that a queue with one customer in service will ever become empty, is exactly the same as the probability that an empty queue will ever again become empty, simply because the identity of the first two rows of the transition matrix shows that a jump from state zero to any other state has the same probability as a jump from state one to that state.

For example, a jump from state zero to state seven represents seven arrivals and has probability, a_7 and a jump from state one to state seven also represents seven arrivals, since the customer being served is discharged during the time unit. This means that the probability of the queue in state one becoming eventually empty is the same as that for the queue in state zero, or $\zeta_1 = \zeta$. Therefore, with state zero transient, $\zeta < 1$ satisfies

$$\zeta = a_0 + \sum_{y=1}^{\infty} a_y \zeta^y. \tag{10}$$

In the other case, where zero is recurrent, $\zeta = 1$ clearly also satisfies this equation. Thus in every case ζ is a root of the equation

$$s = \alpha(s),$$

where $\alpha(s)$ is the probability generating function of the arrival distribution:

$$\alpha(s) = \sum_{j=0}^{\infty} a_j s^j.$$

Referring to Section 1.13, it is easy to see that (except for the special cases given in that section, which will be discussed below) every state of the queue is transient if and only if $\zeta < 1$, which will happen if and only if the mean

$$\lambda = \sum_{j=1}^{\infty} j a_j$$

of the arrival distribution is > 1; every state of the queue is recurrent if and only if $\zeta = 1$, which will happen if and only if $\lambda \le 1$.

This result is intuitively clear, since a mean arrival rate greater than one means that arrivals occur more frequently, on the average, than service terminations, so that the system is overloaded and must gradually increase in size, whatever local fluctuations may be encountered.

Here now are the special cases and their significance:

Case I: $a_0 = 1$. There are never any arrivals to the system, state zero is absorbing, and the chain is reducible.

Case II: $a_1 = 1$. Whatever state the chain begins in is absorbing, and once again the chain is reducible.

Case III: $a_0 + a_1 = 1$. Regardless of the initial state, the queue gradually descends to states zero and one and remains there. The chain is reducible.

Thus it is necessary to assume that $\lambda \leq 1$ in order to find the stationary distribution π_x and its probability generating function $\phi(s)$. From the transition matrix the following equations can easily be written down:

$$\pi_0 a_0 + \pi_1 a_0 = \pi_0,$$

$$\pi_0 a_1 + \pi_1 a_1 + \pi_2 a_0 = \pi_1,$$

$$\pi_0 a_2 + \pi_1 a_2 + \pi_2 a_1 + \pi_3 a_0 = \pi_2, \tag{11}$$

$$\vdots \qquad \vdots$$

with the number of terms increasing in each succeeding equation. Using the technique illustrated in Section 3.9, the nth equation is multiplied by s^{n-1}, $n = 1, 2, 3, \ldots$, and the whole set summed, leading to the following result:

$$\phi(s) = \pi_0 \alpha(s) + \pi_1 \alpha(s) + \pi_2 s \alpha(s) + \pi_3 s^2 \alpha(s) + \cdots,$$

which leads to

$$\phi(s) = \frac{(1-s)\pi_0}{1 - s/\alpha(s)}. \tag{12}$$

Letting $s \to 1$ with the aid of L'Hôpital's rule permits evaluation (when $\lambda \leq 1$) of $\pi_0 = 1 - \lambda$, and thus $\phi(s)$ can be written explicitly in terms of the arrival distribution probability generating function $\alpha(s)$:

$$\phi(s) = \frac{\alpha(s)(1-\lambda)(1-s)}{\alpha(s) - s}. \tag{13}$$

This relationship can be used to obtain moments, probabilities, and special cases, similarly to those obtained in Section 3.9. Since queueing theory will be treated in more detail in Chapter 6, these results will be omitted at present.

3.11. The Ehrenfest[†] Chain

A problem in physics—heat exchange—suggests a Markov chain with nonzero elements only on the principal super- and subdiagonals:

$$p_{x,x+1} = 1 - x/k, \qquad p_{x,x-1} = x/k, \qquad (14)$$

where there are $k+1$ states, $x = 0, 1, 2, \ldots, k$. For example, with $k=3$, the transition matrix defining the chain would be

$$\begin{pmatrix} 0 & 1 & 0 & 0 \\ \frac{1}{3} & 0 & \frac{2}{3} & 0 \\ 0 & \frac{2}{3} & 0 & \frac{1}{3} \\ 0 & 0 & 1 & 0 \end{pmatrix}.$$

Although this chain moves always from left to right or right to left, never staying in place, there is a "central tendency," in that the further away the chain is from the middle, the more likely it is to move towards the middle. In the limiting cases, where $x=0$ or $x=k$, there is a probability of one of a step towards the middle, away from the reflecting boundaries.

It is the tendency towards central equilibrium that originally recommended this matrix as a suitable model for heat exchange. In the physical model, the states of the system are equated to the number of molecules in one of two (heat-exchanging) containers, so that k is, for practical purposes, far larger than three, and, in fact, the student interested in the physical interpretation of the Ehrenfest model would do well, in the following discussion, to think of k as, perhaps, 10^{10}, so that, for example, the transition from state one to state zero would have a probability like 10^{-10}, a very small number.

It is easy to see that the Ehrenfest chain is irreducible but periodic, with period two. Nevertheless, like the trivial chain mentioned in Section 3.7, there is a stationary probability vector associated with the system. The

[†] Paul Ehrenfest, Dutch physicist, 1880–1933.

first steps in the analysis are only slightly different from those of Sections 3.9 and 3.10. The Ehrenfest chain matrix leads to the system of equations

$$\pi_0 = \frac{1}{k}\pi_1,$$

$$\pi_x = \left(1 - \frac{x-1}{k}\right)\pi_{x-1} + \frac{x+1}{k}\pi_{x+1}, \qquad x = 1,\ldots,k-1,$$

$$\pi_k = \left(1 - \frac{k-1}{k}\right)\pi_{k-1},$$

and thence to the generating function equation

$$\phi(s) = s\phi(s) - s^k\pi_{k-1} - \pi_k s^{k+1} + \frac{s}{k}\left[\phi(s) - \pi_k s^k\right]$$

$$+ \frac{1}{sk}\left[\phi(s) - \pi_0\right] - \frac{s}{k}\sum_{j=1}^{k-1} js^{j-1}\pi_{j-1} + \frac{1}{sk}\sum_{j=1}^{k-1} j\pi_{j+1}s^{j+1}.$$

The last two terms do not seem to fit into any simple form involving the probability generating function itself, but a moment's reflection leads to the suspicion that they are similar to mean (expected) values and might be obtained by differentiation of $\phi(s)$. This indeed is the case; here is a *differential equation* in $\phi(s)$. Writing out the (finite) series involved, the student will be able to show that

$$\sum_{j=1}^{k-1} j\pi_{j+1}s^{j+1} = s^2\frac{d}{ds}\left(\frac{\phi(s) - \pi_0}{s}\right)$$

and

$$\sum_{j=1}^{k-1} j\pi_{j-1}s^{j-1} = \frac{d}{ds}\left[s\phi(s)\right] - ks^{k-1}\pi_{k-1} - (k+1)\pi_k s^k.$$

Substituting these values, and simplifying considerably, the following differential equation is obtained:

$$\phi(s) = \frac{1+s}{k}\phi'(s),$$

which can be solved by elementary methods. The result is

$$\phi(s)=\left(\tfrac{1}{2}\right)^k (1+s)^k.$$

This probability generating function is binomial with parameter $p=\tfrac{1}{2}$. Thus the probabilities π_x are given by

$$\pi_x = \binom{k}{x}\left(\tfrac{1}{2}\right)^x\left(\tfrac{1}{2}\right)^{k-x} = \binom{k}{x}2^{-k}, \qquad x=0,1,\ldots,k,$$

a standard binomial distribution. (Keeping in mind the large value of k suggested by the physical model, the student familiar with the central limit theorem will easily understand how this binomial result can be closely approximated by a normal distribution in practical circumstances.)

From the point of view of Markov chain theory, the example of a nontrivial periodic chain with a stationary probability vector is important. The example is also important as a technical exercise in finding a probability generating function by means of a differential equation. Finally, the Ehrenfest chain shows by example that not every finite ergodic chain is regular: The condition of aperiodicity must be added.

3.12. Branching Chains

The Markov chains studied up to this point have all had a certain simplicity: It has been possible to proceed from the model to an explicit form for the transition matrix, and from the matrix to the stationary distribution. These calculations have permitted an easy classification of states with respect to parameter values.

There are, however, certain Markov chains of importance for which even the first step—an explicit form for the matrix—is difficult, so that the analysis is more roundabout. *Branching chains* are of this type.

The idea of a branching chain came originally from the study of the *extinction of surnames*, i.e., whether or not one of the generations following a given individual would consist entirely of married females, or contain no children. Subsequently, several diverse applications of the theory have been discovered, notably in genetics and nuclear physics, where *individuals* can, with known probability, give rise to *offspring*, who in turn can give rise to further offspring of succeeding *generations*.

The key probabilistic assumption is that each individual, of whatever generation, has a probability b_x, $x = 0, 1, 2, \ldots$, of producing x new individuals. Let $\beta(s) = \Sigma b_j s^j$ be the probability generating function of this *birth distribution*, and let $B = \beta'(1)$ be its mean.

The Markov chain is defined by letting X_n represent the number of individuals in the nth generation, where the original ancestor forms the zeroth generation. There is no loss of generality in assuming that $X_0 = 1$, since an initial generation of j individuals would evolve like j branching chains, each with single ancestors and pooled offspring.

Calculation of the transition matrix begins smoothly enough. Since the zeroth state is absorbing, the first row consists of the vector $(1, 0, 0, \ldots)$. The second row represents the probabilities of transitions from one member in a generation, and thus, by definition, is the birth distribution $(b_0, b_1, b_2, \ldots)$. In the third row, the probability of transition from state two to state zero is simply b_0^2; from state two to state one, $2b_0 b_1$; from state two to state two, $b_1^2 + 2b_0 b_1$ (since the two members of the subsequent generation might or might not have the same ancestor), and so forth. But, beginning with three or more members in the anterior generation, the calculations become rather heavy, owing to the numerous possibilities of ancestor–offspring combinations, and it is natural to look for some underlying principle.

The transition probability

$$p_{xy} = P(X_{n+1} = y \mid X_n = x)$$

is conditional upon an nth generation of x individuals. Let $Z_1, Z_2, \ldots, Z_x$ be the number of offspring of these x individuals, with, by hypothesis, the distribution b_x for each of the x independent random variables. Then the $(n+1)$st generation, which has size $\Sigma_{j=1}^x Z_j$, will have distribution b_y^{x*} and probability generating function $[\beta(s)]^x$. Thus

$$p_{xy} = P(X_{n+1} = y \mid X_n = x) = \text{coefficient of } s^y \text{ in} \atop \text{the expansion of } [\beta(s)]^x. \tag{15}$$

It is easy to check this result against the values computed above for the transitions from state two and to see how one might write out the next row systematically. However, unless $\beta(s)$ is particularly simple, this does not greatly facilitate the expression of p_{xy} explicitly in terms of the defining probabilities of the chain, b_x. It is clear, nevertheless, that since all the states communicate with the absorbing state zero, they must be transient, unless it is impossible not to have at least one offspring, $b_0 = 0$. Thus the interest in branching chains lies not so much in the calculation of stationary distributions (since every state has limiting probability zero) as in computing the probabilities for an nth generation of size x, and especially the probability of a transition to state zero, i.e., the *probability of extinction.*

Let the probability of x individuals in the nth generation be denoted by $b_x^{(n)}$, with probability generating function $\beta_n(s)$ and mean B_n:

$$b_x^{(n)} = P(X_n = x), \qquad \beta_n(s) = \Sigma b_x^{(n)} s^x, \qquad B_n = \beta_n'(1),$$

with $b_x^{(1)} = b_x$, $\beta_1(s) = \beta(s)$, $B_1 = B$. Then, by definition of the branching chain,

$$b_x^{(n)} = \sum_{j=0}^{\infty} b_j^{(n-1)} p_{jx}$$

$$= \sum_{j=0}^{\infty} b_j^{(n-1)} \left\{ \text{coefficient of } s^x \text{ in } \left[\beta(s) \right]^j \right\}$$

$$= \text{coefficient of } s^x \text{ in } \sum_{j=0}^{\infty} b_j^{(n-1)} \left[\beta(s) \right]^j. \tag{16}$$

This implies, simply by definition of the probability generating function, that

$$\beta_n(s) = \sum_{j=0}^{\infty} b_j^{(n-1)} \left[\beta(s) \right]^j$$

$$= \beta_{n-1} \left[\beta(s) \right]. \tag{17}$$

The last relationship is fundamental in the study of branching chains.

Differentiating this formula, a recurrence relation for mean values can be easily obtained:

$$B_n = B B_{n-1},$$

from which it follows that

$$B_n = B^n. \tag{18}$$

The similar formula for variances

$$V_n = VM^{n-1} \frac{1-M^n}{1-M}, \qquad M \neq 1,$$

is a little more troublesome, but follows in a perfectly straightforward manner. These calculations are left as an exercise for the student.

Equation (18) shows that the limiting mean generation size $\lim_{n \to \infty} B_n$ can only be $0, 1$, or ∞, depending on whether $B<1$, $B=1$, or $B>1$.

Branching processes form an important example of a more general principle: the formation of probability distributions from known distributions by *nesting* probability generating functions. In the branching process case, the "nest" was formed from the same generating function: $\beta[\beta(s)]$, $\beta\{\beta[\beta(s)]\}$, etc. This process can also be used with different probability generating functions and will clearly always give a probability generating function: $\beta[\alpha(s)]$, $\gamma\{\beta[\alpha(s)]\}$, and so forth. The probabilistic interpretation of the process is important. Consider the sum of identically distributed, independent, random variables $X_1 + X_2 + X_3 + \cdots + X_N$, where N is itself a random variable independent of X_j. This is called a *random sum*, and the process of nesting probability generating functions is equivalent to the random summation of random variables. The proof is essentially the same as in the case of branching processes. Let X_j be independently and identically distributed with probability generating function $\alpha(s)$, let N be independent with probability generating function $\beta(s)$, and let $\phi(s)$ be the probability generating function for the random sum $X_1 + X_2 + \cdots + X_N = S_N$:

$$P(S_N = x) = P\left(\sum_{j=0}^{N} X_j = x \right)$$

$$= \sum_{n=0}^{\infty} P\left(\sum_{j=0}^{n} X_j = x \right) P(N=n).$$

Thus

$$\phi(s) = \sum_{x=0}^{\infty} s^x \sum_{n=0}^{\infty} P\left(\sum_{j=0}^{n} X_j = x \right) P(N=n)$$

$$= \sum_{n=0}^{\infty} P(N=n) [\alpha(s)]^n$$

$$= \beta[\alpha(s)]. \tag{19}$$

Although this procedure has a number of rather important applications, the distributions which are obtained are frequently difficult to compute and have therefore remained rather obscure. In the simple example of a geometric number of Poisson variables, the resulting probability generating function

$$\frac{1-\rho}{1-\rho e^{-\lambda+\lambda s}}$$

resists easy analysis. Many similar examples could be given.

The fundamental generating function equation can be obtained by using the method of marks (Section 2.4). Suppose every individual in the process is independently marked with probability $1-s$. Then the probability generating function $\beta_n(s)$ represents the probability that none of the individuals in the nth generation are marked. This probability can also be written, conditional on x individuals in the first generation, as $[\beta_{n-1}(s)]^x$, and hence

$$\beta_n(s)=\Sigma b_x[\beta_{n-1}(s)]^x,$$

from which Eq. (17) follows.

3.13. Probability of Extinction

In the notation of Section 3.12, let α_n be the probability that the nth generation is empty:

$$\alpha_n=P(X_n=0)=\beta_n(0).$$

It is obvious that $\alpha_1=b_0$ and the basic relationship $\beta_n=\beta_{n-1}(\beta)$ shows that $\alpha_n=\beta(\alpha_{n-1})$. Since the probability generating function β is increasing, it follows that

$$\alpha_1<\alpha_2<\alpha_3<\cdots\leq1.$$

Hence there is a limit α_∞ satisfying

$$\alpha_\infty=\beta(\alpha_\infty).$$

Here a_∞ is interpreted as the probability that the process eventually dies out, or that the "family" eventually becomes "extinct." Equations of this type were discussed in Section 1.13. Let the (possible) root less than one be denoted by σ. Since $\beta(0) < \beta(\sigma) = \sigma$, $\alpha_1 < \sigma$. By induction, all $\alpha_n < \sigma$, and therefore $\alpha_\infty = \sigma$. Thus the value of α_∞ is given by the analysis of Section 1.13.

First, the special cases: (i) If $b_0 = 1$, $\alpha_\infty = 1$; (ii) if $b_1 = 1$, $\alpha_\infty = 0$; (iii) if $b_0 = 1 - b_1$, $\alpha_\infty = 1$. In the general case, $M = 1$ implies that $\alpha_\infty = 1$, and $M > 1$ implies that $\alpha_\infty = \sigma < 1$.

Summing ΣB^n shows that the mean size of the entire progeny is $(1 - B)^{-1}$ when $M < 1$, and infinite when $M \geq 1$.

3.14. The Gambler's Ruin

This Markov chain on the integers $0, 1, \ldots, N$ was introduced in Section 3.2. It models a game in which a gambler can win or lose[†] a unit sum at each stage of a game, with fixed probabilities of win or loss. Thus it is a special case of the random walk discussed in Section 3.2, with

$$\lambda = P(\text{win}) = P(X_n = x | X_{n-1} = x - 1), \qquad x = 2, 3, \ldots, N,$$

$$1 - \lambda = P(\text{loss}) = P(X_n = x - 2 | X_{n-1} = x - 1), \qquad x = 2, 3, \ldots, N,$$

and absorption at the Nth state,

$$P(X_n = N | X_{n-1} = N) = 1,$$

and at the zeroth state,

$$P(X_n = 0 | X_{n-1} = 0) = 1 \qquad (\text{i.e.,} \alpha = 0).$$

Since the two absorbing states can be reached from the intermediate states, it is clear that the states $1, 2, \ldots, N - 1$ are transient.

The problem is to find the probabilities for the eventual absorption into states 0, corresponding to a loss of the initial stake, and N, which could correspond either to winning all the opponent's stake or to a predetermined fortune desired. These probabilities, it is clear intuitively, depend on the gambler's initial stake k, where $k = 1, 2, \ldots, N - 1$.

Let $p(k)$ be the probability of eventual ruin:

$$p(k) = P(X_n = 0 \text{ for some } n | X_0 = k).$$

[†]A trivially more general model permits draws.

Since state x can be reached only from state $x-1$ (by winning) or from state $x+1$ (by losing), the probabilities $p(k)$ satisfy the difference equation

$$p(k)=(1-\lambda)p(k-1)+\lambda p(k+1), \qquad k=1,2,\ldots,N-1, \qquad (20)$$

with boundary conditions $p(0)=1, p(N)=0$. This equation can be rewritten

$$p(k+1)-p(k)=C[p(k)-p(k-1)],$$

where

$$C=\frac{1-\lambda}{\lambda}.$$

Thus, beginning with $p(1)-p(0)$, it is clear that the differences $p(k)-p(k-1)$ form a geometric series with ratio C:

$$p(k+1)-p(k)=C^k[p(1)-1].$$

But since $p(k)=\Sigma_{j=0}^{k-1}[p(j+1)-p(j)]$, the required probability can be written as the sum of a geometric series:

$$p(k)-1=\sum_{j=0}^{k-1}C^j[p(1)-1].$$

In this equation, the remaining value $p(1)$ is evaluated by using the boundary condition $p(N)=0$, yielding

$$p(1)=1-\frac{1-C}{1-C^N}$$

and, finally,

$$p(k)=\frac{[(1-\lambda)/\lambda]^k-[(1-\lambda)/\lambda]^N}{1-[(1-\lambda)/\lambda]^N}, \qquad \lambda\neq\tfrac{1}{2}. \qquad (21)$$

It will be noted that this analysis depends on the assumption that $\lambda\neq\tfrac{1}{2}$. When $\lambda=\tfrac{1}{2}$, the corresponding formula can be obtained by direct arguments analogous, but much simpler, or, alternatively, by using L'Hôpital's rule on the formula already obtained. In either case, the result will be found to be

$$p(k)=1-k/N, \qquad \lambda=\tfrac{1}{2}.$$

Another special case of interest is the game against an infinitely rich adversary, $N \to \infty$. Passing to the limit, again with the use of L'Hôpital's rule, the following values are obtained:

$$p(k)=1, \qquad \text{if } \lambda \leq \tfrac{1}{2},$$

$$p(k)=\left(\frac{1-\lambda}{\lambda}\right)^{k}, \qquad \text{if } \lambda > \tfrac{1}{2}. \tag{22}$$

The student familiar with difference equations will be able to understand the more general problem of determining, in the gambler's ruin formulation, the probability $p_n(k)$ that the gambler will be ruined *at the nth game*, beginning with a stake of k. The double difference equation

$$p_{n+1}(k)=\lambda p_n(k+1)+(1-\lambda)p_n(k-1), \tag{23}$$

with boundary conditions

$$p_n(0)=p_n(N)=0, \qquad n=1,2,3,\ldots,$$

$$p_0(0)=1, \quad p_0(k)=0, \qquad k=1,2,3,\ldots,$$

can be reduced to a single-variable difference equation by defining the generating function[†]

$$\psi_k(s)=\sum_{n=0}^{\infty} p_n(k)s^n,$$

namely,

$$\psi_k(s)=\lambda s \psi_{k+1}(s)+(1-\lambda)s\psi_{k-1}(s),$$

and solved by standard procedures.

[†] Note that this is not a *probability* generating function, inasmuch as $\sum_n p_n(k)=p(k)\neq 1$.

3.15. Probability of Ruin as Probability of Extinction

It is possible to obtain Eqs. (22) from the theory of branching processes. Consider first a gambler who begins with $k=1$ unit as his initial stake. Let the "first generation" be the amount available to him after the first game, i.e., either zero or two units. The probabilities, using the notation of Section 3.12, are clearly $b_0 = 1-\lambda$, $b_2 = \lambda$, and thus the probability of extinction is the smaller root of the equation

$$s=(1-\lambda)+\lambda s^2.$$

The solutions of this quadratic equation are 1 and $(1-\lambda)/\lambda$, so that, in gambler's ruin notation,

$$p(1)=\frac{1-\lambda}{\lambda}.$$

Now suppose that the gambler begins with $k>1$ units as his initial stake. In order to become ruined, the gambler must pass through the different states $k-1, k-2,\ldots,1,0$, and each step has the same independent probability as the probability of ruin with one initial unit; thus

$$p(k)=\left(\frac{1-\lambda}{\lambda}\right)^k.$$

It is left as an exercise for the student to discover where the important condition $\lambda>\frac{1}{2}$ enters this formulation, giving Eqs. (22).

3.16. First-Passage Times

In Section 3.8, the random variable Y was defined to be the number of steps for first passage from a fixed state back to the same state. In this section, the idea is generalized to the number of steps from one fixed state to another fixed state. For states x and y, let Y be the number of steps for first passage from x to y, with

$$f_{xy}(n)=f(n)=P(Y=n).$$

Since a process which goes from x to y in n steps must necessarily reach y for the first time on one of the steps $1,2,\ldots,n$ and then return to y again at the nth step,

$$P(X_n = y \mid X_0 = x) = \sum_{k=1}^{n} P(Y=k)P(X_n = y \mid X_k = y),$$

or, with the notation of Section 3.5,

$$p_{xy}^{(n)} = \sum_{k=1}^{n} f(k)p_{yy}^{(n-k)}. \tag{24}$$

It is possible to calculate the values of $f(n)$ recursively from these equations (letting $p = p_{xy}$, $q = p_{yy}$):

$$f(1) = p,$$

$$f(2) = p^{(2)} - qf(1),$$

$$f(3) = p^{(3)} - qf(2) - q^{(2)}f(1),$$

$$f(4) = p^{(4)} - qf(3) - q^{(2)}f(2) - q^{(3)}f(1),$$

$$\vdots \qquad \qquad \vdots$$

where the upper indices refer to the corresponding powers of the basic matrix. Thus, in order to calculate $f(n)$ systematically, one needs to know two elements from all powers of $\mathbf{P}$, namely, the elements $p_{xy} = p$ and $p_{yy} = q$. When these are known, the calculation can be reduced to a single stage by using probability generating functions. Multiplying the jth equation above by s^j and summing gives an expression for the probability generating function $\phi(s) = \sum s^n f(n)$:

$$\phi(s) = \sum_{j=1}^{\infty} s^j p^{(j)} - \phi(s) \sum_{j=1}^{\infty} s^j q^{(j)}.$$

Defining the functions $p(s)$ and $q(s)$ as follows,

$$p(s) = \sum_{j=1}^{\infty} s^j p^{(j)}, \qquad q(s) = \sum_{j=1}^{\infty} s^j q^{(j)},$$

$\phi(s)$ can be written

$$\phi(s)=\frac{p(s)}{1+q(s)}.\qquad\qquad (25)$$

Example 1. Consider a sequence of Bernoulli trials with $\beta=P(\text{Success})$ and with $X_n=1,2,3,4$, accordingly to the $(n-1)$st and nth trials resulting in SS, SF, FS, FF. Then

$$\mathbf{P}=\begin{pmatrix} \beta & 1-\beta & 0 & 0 \\ 0 & 0 & \beta & 1-\beta \\ \beta & 1-\beta & 0 & 0 \\ 0 & 0 & \beta & 1-\beta \end{pmatrix}$$

and

$$\mathbf{P}^n=\begin{pmatrix} \beta^2 & \beta(1-\beta) & \beta(1-\beta) & (1-\beta)^2 \\ \beta^2 & \beta(1-\beta) & \beta(1-\beta) & (1-\beta)^2 \\ \beta^2 & \beta(1-\beta) & \beta(1-\beta) & (1-\beta)^2 \\ \beta^2 & \beta(1-\beta) & \beta(1-\beta) & (1-\beta)^2 \end{pmatrix},\qquad n=2,3,4,\dots.$$

Suppose first passage from state one to state two is required. Then

$$p(s)=(1-\beta)s+\beta(1-\beta)(s^2+s^3+\cdots)$$

$$=\frac{(1-\beta)s(1-s+s\beta)}{1-s},$$

$$q(s)=\frac{s^2\beta(1-\beta)}{1-s},$$

$$\phi(s)=\frac{(1-\beta)s}{1-s\beta},$$

$$f(n)=(1-\beta)\beta^{n-1},\qquad n=1,2,3,\dots,$$

which is a shifted geometric distribution. It is important to note regarding $p(s)$ and $q(s)$ that, although they are generating functions, they are not probability generating functions and hence do not necessarily pass through the point $(1,1)$. In fact, in this example, both are infinite at the value $s=1$. However, $\phi(s)$ is a probability generating function and $\phi(1)=1$.

The calculation depends, of course, on the two states chosen, $x=1$, $y=2$. In this example, there are 16 different possible choices of state, including those of the form $x=y$. The first-passage distributions obtained are not all geometric (one other.is) and are not all equal. The student should find a few of these generating functions as an exercise.

In the case $x=y$, the first passage (now recurrence time, cf. Section 3.8) is from a state back to the same state, and the general formula is still valid with $p(s)=q(s)$, as may be shown by recapitulating its proof. These first-passage probabilities are connected with the quantities ξ and $f_x(n)$, defined in Section 3.8 in the following way:

$$f_x(n)=f_{xx}(n)=f(n),$$

where $f(n)$ is the distribution from state x back to state x and

$$\xi= \sum_{n=1}^{\infty} f(n).$$

Of course, by definition, the first-passage probabilities from one state back to the same state sum to unity if and only if that state is recurrent. For transient states,

$$\sum_{n=1}^{\infty} f(n)<1.$$

This inequality seems to contradict the basic postulates of probability, that every distribution must sum to unity. However, the contradiction is only an illusion based on a misunderstanding of the values of the random variable. Summed over the positive integers, the distribution is indeed *deficient* or *improper*,[†] but this simply means that the possible first-passage times are not fully described by the positive integers. Indeed, there is another possible result, namely, that the passage never terminates, and this possibility has probability

$$1- \sum_{n=1}^{\infty} f(n).$$

[†] These are the usual, but somewhat misleading, names given to such distributions.

If one wished, this probability might be denoted by some convenient symbol such as $f(\infty)$ or $f(\Omega)$. With a correct accounting for possible values, distributions are not "proper" and "improper"—they are all proper.

Another example of a distribution augmented beyond the positive integers is that of the number of customers passing through a queue during a busy period, which is a first passage from zero to zero. This number may be $1, 2, 3, \ldots$, but it may also happen that the queue never again becomes empty, and that a certain probability must be assigned to the possibility. Here it is perhaps more appropriate to write $f(\infty)$, since one can say with intuitive justification that the number of customers passing through is infinite, although, strictly speaking, the number is not defined, since the queue never becomes empty. A similar example is obtained by considering the total number of offspring in a branching process.

The generating function equation (25) is closely connected with the problem of proving the fundamental relation referred to in Section 3.8:

$$\pi_x \mu(x) = 1.$$

First note that

$$\mu(x) = \sum_{j=1}^{\infty} jf(j) = \phi'(1),$$

where the f and ϕ functions apply to state x. The generating function equation can be written in the form

$$\frac{\phi(s)}{1-\phi(s)} = \sum_{j=1}^{\infty} s^j q^{(j)},$$

since $p(s) = q(s)$ for return to the same state. There is a theorem[†] which states that in this situation

$$\lim_{n \to \infty} q^{(n)} = \frac{\phi(1)}{\phi'(1)},$$

which in this case becomes $1/\mu(x)$. But the limit of the diagonal element in the matrix has already been shown to be π_x. This means that when $q(s)$ has

---

[†] For the exact conditions on the theorem, an outline of the proof, and complete references, see Cox, D. R., and MILLER, H. D. (1965), *The Theory of Stochastic Processes*, Methuen, London, pp. 140–141.

been found for any state x, then $\mu(x)$ can be calculated as

$$\mu(x) = \phi'(1)$$

$$= \frac{d}{ds}\left(\frac{q(s)}{1+q(s)}\right)_{s=1}$$

$$= \frac{q'(1)}{[1+q(1)]^2}. \tag{26}$$

The student can verify this formula for Example 1 above.

Recommendations for Further Study

There are a number of practical applications given as problems in Chapter 4 of LINDLEY, D. V., (1965), *Introduction to Probability and Statistics from a Bayesian Viewpoint, Part 1*, Cambridge University Press, Cambridge.

3.17. Problems[†]

1. Let X_n be a Markov chain with state space $0, 1, 2$, with initial vector (a_0, a_1, a_2) and transition matrix $\{p_{xy}\}$. Find (i) $P(X_0 = 0, X_1 = 1, X_2 = 1)$, (ii) $P(X_1 = 1, X_2 = 1 | X_0 = 0)$, (iii) $P(X_n = 1 | X_{n-2} = 0)$, (iv) $P(X_2 = X_0)$.

2. A fair coin is tossed until three consecutive heads occur. Let $X_n = x$ if at the nth trial the last tail occurred at the $(n-x)$th trial, $x = 0, 1, \ldots, n$, i.e., X_n denotes the length of the string of heads ending at the nth trial. Write down the transition matrix.

   $$Ans. \quad \begin{pmatrix} \frac{1}{2} & \frac{1}{2} & 0 & 0 \\ \frac{1}{2} & 0 & \frac{1}{2} & 0 \\ \frac{1}{2} & 0 & 0 & \frac{1}{2} \\ 0 & 0 & 0 & 1 \end{pmatrix}$$

3. Two green balls and two yellow balls are placed in two boxes so that each box contains two balls. At each step one ball is selected at random from each box, and the two exchanged. Let X_0 denote the number of yellow balls initially in the first box. For $n = 1, 2, 3, \ldots$, let X_n denote the number of yellow balls in the first box after n exchanges have taken place. (i) Find the transition matrix and the two-step transition matrix. (ii) Show that $\lim_{n \to \infty} X_n = X_0$. (iii) Find the one-step matrix beginning with N green and N yellow balls.

[†]Problems 17 and 27 are taken from Lindley (1965), with kind permission of the publishers.

4. There are two green balls in box A and three red balls in box B. At each step a ball is selected at random from each box and the two are exchanged. Let the state of the system at time n be the number of red balls in box A after the nth exchange. (i) Find the transition matrix. (ii) What is the probability that there are two red balls in box A after three steps? (iii) What is the long-run probability that there are two red balls in box A? *Ans.* (ii) $\frac{5}{18}$, (iii) $\frac{3}{10}$

5. A number X_0 is chosen at random from the integers $1,2,3,4,5$. For $n=1,2,3,...$ a value of X_n is chosen at random from the integers $1,2,...,X_{n-1}$. Find the one-step and two-step transition matrices.

6. In a sequence of independent throws of a fair die, let X_n be the largest number appearing during the first n throws. Find the one-step and two-step transition matrices.

7. Four children throw a ball to one another. A child with the ball is equally likely to throw it to each of the other three children. Find the one-step and two-step transition matrices.

8. (A famous problem). A and B have agreed to play a series of games until one of them has won five games. They are equally strong players and no draws are allowed. Owing to circumstances beyond their control, they are compelled to stop at a point where A has won four and B has won three games. How should the stakes be divided? The present and possible future states of the system are four: A has four games, B three; A has four and B also has four; A has won; B has won. Write down the transition matrix and solve the problem.

9. Consider a two-state Markov chain with $p_{00}=\lambda$, $p_{11}=\mu$. A new chain is constructed by calling the pair 01 state 0 and the pair 10 state 1, ignoring 00 and 11. Only nonoverlapping pairs are considered. Show that for the new chain

$$p_{00}=p_{11}=(1-\lambda-\mu)^{-1}.$$

10. In independent throws of a coin which has probability p of falling heads, let X_n be the number of heads in the first n throws. Find the one-step and two-step transition matrices.

11. Let X_n be a two-state Markov chain over $1,2$. Find (i) $P(X_1=1|X_0=1$ and $X_2=1)$ and (ii) $P(X_1 \neq X_2)$.

12. A particle moves on a circle through points $0,1,2,3,4$ in clockwise order. At each step it has probability p of moving to the next point clockwise and probability $1-p$ of moving to the next point counterclockwise. Find the transition matrix.

13. A rat is being trained to run a maze, and each attempt is considered to be a success or a failure. Suppose the sequence of trials forms a Markov chain with transition matrix

$$\begin{matrix} S \\ F \end{matrix} \begin{pmatrix} \frac{2}{3} & \frac{1}{3} \\ \frac{1}{2} & \frac{1}{2} \end{pmatrix}$$

and that on the first trial the rat is equally likely to succeed or fail. The rat is considered to be trained if it achieves three consecutive successes. Find the probability that the rat is not trained after 10 trials.

14. Test the following matrices for regularity.

$$(i) \quad \begin{pmatrix} 0 & 1 \\ \frac{1}{2} & \frac{1}{2} \end{pmatrix}, \qquad (ii) \quad \begin{pmatrix} 1 & 0 \\ \frac{1}{2} & \frac{1}{2} \end{pmatrix}, \qquad (iii) \quad \begin{pmatrix} \frac{1}{2} & \frac{1}{2} \\ 0 & 1 \end{pmatrix},$$

$$(iv) \quad \begin{pmatrix} \frac{1}{2} & \frac{1}{4} & \frac{1}{4} \\ 0 & 1 & 0 \\ \frac{1}{2} & \frac{1}{2} & 0 \end{pmatrix}, \qquad (v) \quad \begin{pmatrix} 0 & 0 & 1 \\ \frac{1}{2} & \frac{1}{4} & \frac{1}{4} \\ 0 & 1 & 0 \end{pmatrix}.$$

15. A Markov chain has transition matrix

$$P = \begin{pmatrix} 0 & 1 & 0 \\ 1-p & 0 & p \\ 0 & 1 & 0 \end{pmatrix}.$$

Find P^n.

16. A Markov chain has the transition matrix

$$\begin{pmatrix} 1-p & p & 0 \\ 0 & 1-p & p \\ p & 0 & 1-p \end{pmatrix}.$$

Show that the stationary vector is rectangular.

17. A Markov chain has the following transition matrix:

$$\begin{pmatrix} \frac{1}{2} & \frac{1}{2} & 0 & 0 \\ \frac{1}{4} & \frac{1}{2} & \frac{1}{4} & 0 \\ 0 & \frac{1}{4} & \frac{1}{2} & \frac{1}{4} \\ 0 & 0 & \frac{1}{2} & \frac{1}{2} \end{pmatrix}.$$

(i) Find the stationary vector. (ii) Write down the mean recurrence times.
                                                                Ans. (ii) $(3,6,6,3)$

18. A production line produces three variants A, B, and C of a basic design. For a "balanced" line, the following rules are observed: Two identical variants never follow one another and each B must be followed by a C. (i) What is the probability that after an A, the next A is separated by only one unit? (ii) Given a production schedule which specifies the proportion of each variant required, how could one find a rule which would yield this output in a balanced production line? (iii) Suppose the demand was for 40% variant A, 10% variant B, and 50% variant C. How could this be achieved?

19. Consider a Markov chain with state space $(1,2,4)$, initial vector $(\frac{1}{4},\frac{1}{4},\frac{1}{2})$, and transition matrix

$$\begin{pmatrix} \frac{1}{5} & \frac{2}{5} & \frac{2}{5} \\ 0 & \frac{1}{2} & \frac{1}{2} \\ \frac{1}{4} & \frac{3}{4} & 0 \end{pmatrix}.$$

(i) Find $P(X_0=1, X_1=1, X_3=1)$. (ii) Compute $p_{21}^{(2)}$.

20. A Markov chain with the non-negative integers as its state space has transition matrix

$$\begin{pmatrix} 0 & 1 & 0 & 0 & 0 & \cdots \\ \frac{1}{2} & 0 & \frac{1}{2} & 0 & 0 & \cdots \\ 0 & \frac{3}{4} & 0 & \frac{1}{4} & 0 & \cdots \\ 0 & 0 & \frac{7}{8} & 0 & \frac{1}{8} & \cdots \\ 0 & 0 & 0 & \frac{15}{16} & 0 & \cdots \\ \vdots & \vdots & \vdots & \vdots & \vdots & \end{pmatrix}.$$

(i) Write out the transition probabilities, i.e., fill in the blanks in the following formula: $P(X_{n+1}= \quad |X_n= \quad)= \quad$. (ii) Let the stationary vector be $(\pi_1, \pi_2, \pi_3 \ldots)$. Write a difference equation giving π_x in terms of π_{x-1} and π_{x+1}. (iii) Express π_4 in terms of π_0. (iv) Find $P(X_{n+2}=2|X_n=2)$. (v) Let $\phi(s)=\Sigma\pi_j s^j$. Show that $\phi(s)=P(s)\phi(\frac{1}{2}s)$, where $P(s)$ is a polynomial in s, and find $P(s)$. (vi) Let μ be the mean value of the stationary vector, and let ν be the slope of $\phi(s)$ at the value $x=\frac{1}{2}$. Prove that $2\mu=2\nu+1$.

21. Find the stationary probability vector for

(i) $\begin{pmatrix} 0 & \frac{1}{2} & \frac{1}{2} \\ \frac{1}{3} & \frac{2}{3} & 0 \\ 0 & 1 & 0 \end{pmatrix}$, (ii) $\begin{pmatrix} 0 & 1 & 0 \\ \frac{1}{2} & 0 & \frac{1}{2} \\ \frac{1}{2} & \frac{1}{4} & \frac{1}{4} \end{pmatrix}$, (iii) $\begin{pmatrix} 0 & 1 & 0 \\ 0 & 0 & 1 \\ \frac{1}{2} & \frac{1}{2} & 0 \end{pmatrix}$.

Ans. (i) $(\frac{2}{9},\frac{6}{9},\frac{1}{9})$, (ii) $(\frac{5}{15},\frac{6}{15},\frac{4}{15})$, (iii) $(\frac{1}{5},\frac{2}{5},\frac{2}{5})$

22. Classify the states of the Markov chains with transition matrices

(i) $\begin{pmatrix} \frac{1}{4} & \frac{1}{2} & 0 & \frac{1}{4} \\ \frac{1}{5} & 0 & \frac{1}{3} & \frac{7}{15} \\ 0 & \frac{2}{3} & \frac{1}{3} & 0 \\ \frac{1}{4} & \frac{1}{4} & \frac{1}{4} & \frac{1}{4} \end{pmatrix}$, (ii) $\begin{pmatrix} \frac{1}{3} & \frac{2}{3} & 0 & 0 \\ \frac{3}{4} & \frac{1}{8} & \frac{1}{8} & 0 \\ 0 & 0 & \frac{1}{2} & \frac{1}{2} \\ 0 & 0 & \frac{1}{3} & \frac{2}{3} \end{pmatrix}$

23. Two boys and two girls are throwing a ball. Each boy throws the ball to the other boy with probability $\frac{1}{2}$ and to each girl with probability $\frac{1}{4}$. Each girl

throws the ball to each boy with probability $\frac{1}{2}$ and never to the other girl. Find the long-range probability that each has the ball.　　*Ans.* $(\frac{1}{3}, \frac{1}{3}, \frac{1}{6}, \frac{1}{6})$

24. Let X_n, $n=0,1,2,\ldots,$ be a two-state Markov chain ($p_{00}=\lambda, p_{11}=\mu$), which starts in state 0. Find the probability that the first return to state 0 is at the nth step.　　*Ans.* The p.g.f. $\phi(s)$ is $(1-\mu s)^{-1}(\lambda s+s^2-\lambda s^2-\mu s^2)$.

25. A Markov chain has matrix

$$\begin{pmatrix} 0 & 1 & 0 \\ 1-p & 0 & p \\ 0 & 1 & 0 \end{pmatrix}.$$

(i) Find the n-step transition matrix. (ii) Describe the limiting behavior of the chain. (iii) Find the stationary vector and comment on its meaning.
　　Ans. (iii) $(\frac{1}{2}(1-p), \frac{1}{2}, \frac{1}{2} p)$

26. A fair tetrahedral die with $1,2,3,4$ on its faces is thrown repeatedly, and $X_n=x$ if x is the highest value obtained (on the bottom) in n throws. (i) Write the transition matrix for the Markov chain. (ii) For fixed x and y, find

$$\phi(s)=\sum_j f_{xy}(j)s^j,$$

where $f_{xy}(n)$ is the probability of first transition from x to y on the nth step. (iii) Consider $\phi(1)$ for $y=1,2,3,4$ and explain why $\phi(1)=1$ holds only for $y=4$.

27. A certain kind of nuclear particle splits into 0, 1, or 2 particles with probabilities $\frac{1}{4}$, $\frac{1}{2}$, and $\frac{1}{4}$, respectively, and then dies. The individual particles act independently of each other. Given a particle, let X_1, X_2, and X_3 denote the number of particles in the first, second, and third generations. Find (i) $P(X_2>0)$, (ii) $P(X_1=1|X_2=1)$, (iii) $P(X_3=0)$.

28. In an Ehrenfest chain, show that if the distribution of X_0 is $\binom{k}{x} 2^{-k}$, so is the distribution of X_1. For an Ehrenfest chain with $k=3$, find the distribution of X_1, X_2, and X_3 if X_0 is rectangular.

29. For a branching chain, calculate the probability of extinction for (i) $b_0=\frac{1}{4}, b_2=\frac{3}{4}$, (ii) $b_0=\frac{1}{4}, b_1=\frac{1}{2}, b_2=\frac{1}{4}$, (iii) $b_0=\frac{1}{6}, b_1=\frac{1}{2}, b_2=\frac{1}{3}$, (iv) $b_0=b_3=\frac{1}{2}$.
　　Ans. (i) $\frac{1}{3}$, (ii) 1, (iii) $\frac{1}{2}$, (iv) $-\frac{1}{2}+\frac{1}{2}5^{1/2}$

30. In a certain branching process, the probability of n offspring from one ancestor is geometric with parameter ρ. (i) Find the range of values for ρ which will make the process die out with probability one. (ii) For ρ outside this range, find the probability of extinction. (iii) If ρ is chosen so that the probability of the process never dying out is 0.999, what is the probability that an individual will have no offspring? (iv) Answer the first three parts with the modified assumption that the probability of no offspring is the sum of the first two geometric probabilities, the probability of one offspring is the sum of the next two geometric probabilities, and so forth.

31. Suppose each man has exactly three children, with equal probability that each child is a boy or a girl. Consider the branching chain in which X_n is the number of males in the nth generation. (i) Find the probability that the male line of a given man eventually becomes extinct. (ii) If a given man has two boys and a girl, find the probability that his male line eventually becomes extinct.

32. Consider a branching chain with initial population size N and probability generating function $1-p+ps$. Find the probability distribution of the step x at which the population becomes extinct.

33. At time zero, a blood culture starts with one red cell. At the end of one minute, the red cell dies and is replaced by one of the following combinations: two red cells with probability $\frac{1}{4}$, one red and one white cell with probability $\frac{2}{3}$, two white cells with probability $\frac{1}{12}$. Each red cell lives for one minute and gives birth to offspring in the same way; each white cell lives for one minute and dies without reproducing. Assume the cells behave independently. (i) At time $n+\frac{1}{2}$ minutes after the culture began, what is the probability that no white cells have yet appeared? (ii) What is the probability that the entire culture dies out eventually?

$$Ans. \text{ (i)}(\tfrac{1}{4})^{2^{n+1}-1}, \text{ (ii) } \tfrac{1}{3}$$

34. Bets of $1 each are made on the tosses of a fair coin, with the policy to stop when the winnings reach $10 or the losses reach $20. Find (i) the probability of losing, (ii) the expected loss, (iii) the expected number of bets.

35. A Markov transition matrix is said to be *doubly stochastic* if the columns sum to unity. Show that such a chain has a rectangular stationary vector, if it is irreducible, aperiodic and finite.

36. Consider a Markov chain with state space $1,2,\ldots,c+d$, where c and d are positive integers. Starting from any one of the first c states, transition is equally likely to any of the last d states; starting from any one of the last d states, transition is equally likely to any one of the first c states. (i) Show that the chain is irreducible. (ii) Find the stationary vector.

37. Let X_n be a branching process with probability generating function $\phi(s)$. Let Y_n denote the total number of individuals in the first n generations, i.e.,

$$Y_n = X_0 + X_2 + \cdots + X_n, \qquad n=0,1,2,\ldots, \qquad X_0=1.$$

Let $\psi_n(s)$ be the probability generating function of Y_n. Prove that

$$\psi_{n+1}(s)=s\phi[\psi_n(s)], \qquad n=0,1,2,\ldots .$$

38. Suppose all elements in column y of a Markov transition matrix are equal and nonzero, except that $p_{yy}=0$. Show that the first-passage distribution from x to y is geometric.

39. In the gambler's ruin formulation of Section 3.14, let $q(k)$ denote the probability that the duration of the game will be k stages. (i) Show that $q(k)$ satisfies

$$q(k)=\lambda q(k+1)+(1-\lambda)q(k-1)+1.$$

(ii) With boundary conditions $q(0)=q(N)=0$, show that

$$q(k)=\frac{k}{1-2\lambda}-\frac{N}{1-2\lambda}\frac{1-[(1-\lambda)/\lambda]^{k}}{1-[(1-\lambda)/\lambda]^{N}}.$$

40. Solve sequentially the random walk equations with step probabilities $p_{x-1,x}, p_{xx}, p_{x+1,x}$ to obtain π_x in terms of π_0.

41. In a Markov chain with transition matrix

$$\begin{pmatrix} 0 & \frac{1}{2} & \frac{1}{2} \\ \frac{1}{2} & 0 & \frac{1}{2} \\ \frac{1}{2} & \frac{1}{2} & 0 \end{pmatrix}$$

and states $1,2,3$, show that the first-passage distribution $f_{12}(n)=2^{-n}$. [Show that for odd n, $p^{(n)}=(2^{n}+1)/2^{n}\cdot3$, $q^{(n)}=(2^{n}-2)/2^{n}\cdot3$, and for even n, $p^{(n)}=(2^{n}-1)/2^{n}\cdot3$, $q^{(n)}=(2^{n}+2)/2^{n}\cdot3$.]

42. In the example of Section 3.16, find all of the first-passage probability generating functions.

43. Define a Markov chain based on the fish-netting experiment of Section 2.1, classify the states, and write down the transition matrix.

44. Referring to Eq. (19), show that $E(S_N)=E(X_j)E(N)$.

45. For the Markov chain defined in Problem 25, consider first passage from state 1 to state 3. Find $p(s)$, $q(s)$, and show that

$$\phi(s)=\frac{ps^2}{1-(1-p)s^2},$$

and thus that the first-passage distribution is geometric over the even integers with parameter $1-p$.

4

Continuous Probability Distributions

4.1. Examples

There are many random variables whose range[†] consists of the positive real line $(0, \infty)$ or a portion of the real line.

Example 1. A coin is tossed. Let $X =$ angle formed by the axis of the coin and the edge of the table. Range: $(0, \frac{1}{2}\pi)$.

Example 2. A die is thrown. Let $X =$ duration of time before the die comes to rest. Range: $(0, \infty)$.

Example 3. Radioactive particles are recorded. Let $X =$ duration of time between the recording of two consecutive particles. The range is $(0, \infty)$, or for counters with a *dead time* of length t_0 following each recording, during which the counter is inoperative, the range is (t_0, ∞).

Example 4. A measurement of length is made; $X =$ error in measurement. The range is $(0, \infty)$, or if the error is "signed," the range is $(-\infty, \infty)$.

[†]Up to now no particular name has been attached to the set of values for which a random variable has a positive probability, and indeed there is no standard word in the literature for this important concept. Most authors skirt the question, as the first three chapters of this book have done, by saying that the random variable is "over" or "on" some set of values. Feller attributes his reluctance to use the correct mathematical term "range" (for values assumed by a function) to a statistical definition of "range" as the difference between the largest and smallest items in a sample. This does not seem to be a good enough reason, and from this point onwards, the aspect of a random variable as a function on the sample space will be emphasized by introducing the term "*range*" for the possible values of a random variable.

Example 5. An individual is selected; X = age of the individual. Range: $(0, \infty)$.

Example 6. A telephone call is selected; X = duration of the call. Range: $(0, \infty)$.

Example 7. The first accident in a given year is noted; X = time since the beginning of the year. Range: $(0, 1)$, with years as units.

Example 8. A sample of substance is obtained and a test performed to determine one of its properties; X = difference between the result and the "standard" result. In this case the range will depend on the units used to measure the property. If, however, X = ratio of the result to the "standard" result, the range would be $(0, \infty)$.

In certain cases, such as those involving money, the discrete units, although appropriate by definition, are so small that the range can more conveniently considered to be continuous.

Example 9. A year and a locality are specified; X = maximum temperature. Range: $(0, \infty)$.

Example 10. A manufactured item is selected from a batch; X = strength of item. The range depends on the nature of the testing machinery. See Examples 14–19.

Example 11. A day is chosen at random; X = value of exports on that day. Approximate range: $(0, \infty)$.

Example 12. A manufactured product is subject to quality control; X = time required for the inspection. The range in this case might well be truncated away from zero by the fixed duration of some step in the inspection process.

In addition to wholly continuous random variables, there are certain random variables having ranges which are partly discrete and partly continuous. These are called *mixed* random variables.

Example 13. A person approaches a postal counter; X = time needed to wait for service. Range: 0 and $(0, \infty)$, the first value corresponding to those situations where the counter is empty on arrival. *Waiting time* distributions furnish an important example of mixed random variables, and are treated in Chapter 6.

Example 14. A line is drawn on a sheet of paper which contains a circle; $X=$ length of line cut off by the circle. Range: 0, $(0, D)$, where D is the diameter of the circle. The initial value corresponds to all those lines which miss the circle entirely.

Example 15. An accident is selected, for which compensation may or may not be paid; $X=$ value of compensation. Range: 0, $(0, \infty)$, where the first value corresponds to accidents with no compensation; in practice, those with a loss lower than the insurance-deductible limit. Similarly, the upper limit would, in actual cases, often be replaced by some maximum amount of coverage.

Example 16. A day is chosen and the amount of rainfall recorded; $X=$ amount of rainfall. Range: 0, $(0, \infty)$. The discrete element in the range corresponds to completely dry days.

Example 17. A car is chosen; $X=$ speed of this car. Range: 0, $(0, \infty)$, where the discrete value corresponds to stationary cars.

Example 18. A person and a day are selected; $X=$ number of cigarettes or quantity of beer consumed. What range would be most appropriate for modeling?

In the above examples it might be asked why the range includes the special value zero rather than simply assigning the probability required to the zero in the continuous portion of the range. The answer, which will become clearer in the course of the chapter, is that each number in a continuous range must have probability zero. Indeed, one definition of a continuous random variable X is the equation

$$P(X=x)=0 \qquad \text{for all } x.$$

Consider Example 17. It might happen that half of all cars are stationary, with the remaining car speeds continuously distributed over the interval $(0, \infty)$. The probability model for this situation would assign $p_0 = \frac{1}{2}$, with the remaining probability distributed like a continuous random variable.

Similarly, a fixed proportion of accidents would have no compensation, a fixed proportion of lines fail to touch the circle, and a fixed proportion of days have no rainfall.

The types of random variables considered in this book are three: discrete, ("wholly") continuous, and mixed (discrete and continuous). Much of the present chapter is devoted to describing the probabilistic formulation needed for the latter two types. It is a little more difficult to define probability, moments, generating functions, convolutions, conditional probability, and similar concepts from Chapters 1 and 2 for continuous and mixed random variables.

4.2. Probability Density Functions

The probability structure governing a continuous random variable X can be given by a continuous ("probability density") function $f(x)$ defined over the range of the random variable, with the integral between two values x and y representing the probability that X lies between these values:

$$\int_x^y f(u)\,du = P(x < X < y).$$

In order to qualify as a *probability density function*, f must be positive (or zero outside the range of X), and must normalize to unity, i.e.,

$$\int_{-\infty}^{\infty} f(u)\,du = \int_a^b f(u)\,du = 1,$$

where (a, b) is the range of X. Continuous random variables treated in this book do not have negative values, and therefore the range will habitually be $(0, \infty)$ or a portion thereof. With the convention that f is identically zero outside this range, the usual limits of integration will be $(0, \infty)$, although in a few cases finite limits will be needed. The formulation of a continuous random variable is quite analogous to that of a discrete random variable, with the range $(0, \infty)$ replacing the range $0, 1, 2, \ldots$, integration replacing summation, the function $f(x)$ replacing the function p_x, etc.

It is, however, not possible to use the basic notation of discrete random variables $P(X=x)$, since, as mentioned in Section 4.1, each of a continuously infinite number of values must have probability zero. The difficulty is not hard to overcome, however, since by using the distribution function, both discrete and continuous variables can be represented in identical probability statements,

$$P(X<x)=P_x, \qquad x=1,2,3,\ldots, \text{ where } P_x = \sum_{j=0}^{x-1} p_j$$

in the discrete case, and the virtually identical

$$P(X<x)=F(x), \qquad 0<x<\infty, \text{ where } F(x)=\int_0^x f(u)\,du$$

in the continuous case. The difference in notation between P_x and $F(x)$ reflects a traditional tendency to write discrete variables as subscripts and continuous variables in parentheses.

The Exponential Distribution $f(x)=\lambda e^{-\lambda x}$, $0<x<\infty$, $0<\lambda<\infty$, $F(x)$ $=1-e^{-\lambda x}$ (sometimes called *negative exponential*; both expressions are used in this book).

The Gamma Distribution $f(x)=\lambda^k e^{-\lambda x}x^{k-1}/\Gamma(k), 0<x<\infty, 0<\lambda<\infty,$ $k=1,2,3,\ldots,$

$$F(x)= \frac{1}{\Gamma(k)} \int_0^x \lambda^k e^{-\lambda u}u^{k-1}\,du$$

$$= \frac{1}{\Gamma(k)} \int_0^{\lambda x} e^{-v}v^{k-1}\,dv.$$

The integral is clearly related to the integral defining the gamma function (Section 1.9) and is called an *incomplete gamma function*. The incomplete gamma function will be discussed in detail in Section 4.5.

The student will note that the exponential distribution results from setting $k=1$ in the gamma distribution.

Many of the concepts defined in Chapters 1 and 2 can be extended to the case of continuous random variables by substituting integration for summation.

The events considered here[†] are real intervals (a,b), with probability $\int_a^b f(u)\,du$.

[†]More general definitions are given in theoretical probability.

The mean and variance of a random variable with probability density function $f(x)$ are defined to be

$$\text{mean} = m = \int_0^\infty xf(x)dx,$$

$$\text{variance} = v = \int_0^\infty x^2 f(x)dx - m^2.$$

The expected value of any function $g(x)$ is defined to be

$$E(g(X)) = \int_0^\infty g(x)f(x)dx,$$

with $m = E(X)$. These various definitions depend, naturally, on the integral being finite. If the integral diverges, the various quantities do not exist. A well-known example of a distribution without a finite mean is the *Cauchy distribution*,

$$f(x) = \frac{1}{\pi}\frac{1}{1+x^2}, \qquad -\infty < x < \infty,$$

which is constructed by using a quadratic denominator, just as the corresponding discrete Example 2, Section 1.6.

The *tail* of the distribution is defined to be $1 - F(x)$. In Section 4.8, the standard terminology *distribution function* $F(x)$ will be introduced, combining discrete, continuous, and mixed cases into a single format.

In place of the probability generating function, it is more convenient to use a slightly modified form called the *Laplace transform*:

$$\phi(s) = \int_0^\infty e^{-sx}f(x)dx.$$

The Laplace transform will be treated more generally in Section 4.10.

Bivariate and conditional continuous distributions can similarly be defined, using the joint density function $f(x, y)$. If the marginal (or unconditional) distributions are

$$g(x) = \int_0^\infty f(x, y)dy \quad \text{and} \quad h(y) = \int_0^\infty f(x, y)dx,$$

then the conditional densities are given by

$$\frac{f(x, y)}{g(x)} \quad \text{and} \quad \frac{f(x, y)}{h(y)}.$$

It is usually not convenient to define new symbols to denote the cumulative forms, moments, and other quantities that can be associated with a bivariate distribution. These will be developed as the need arises, as in the following example.

An important property of the exponential distribution is that if truncated, it remains similar to itself.[†] That is, if a variable X has density function $\lambda e^{-\lambda x}$, then so does $X-k$ under the additional condition that $X>k$. To see this, consider the cumulative forms:

$$P(X<x|X\geq k)=\frac{P(X<x, X>k)}{P(X>k)}$$

$$=\frac{P(X<x)-P(X<k)}{P(X>k)}$$

$$=\frac{1-e^{-\lambda x}-(1-e^{-\lambda k})}{e^{-\lambda k}}$$

$$=1-e^{-\lambda(x-k)},$$

and with a change of origin from 0 to k, this becomes the cumulative form of the exponential distribution.

Calculations of integrals with respect to a continuous density can often be greatly facilitated by using the fact that the given density is normalized. Thus, since

$$\int_0^\infty \{\lambda^k e^{-\lambda x}x^{k-1}/\Gamma(k)\}dx=1$$

is given, integrals of the form

$$\int_0^\infty e^{-\lambda x}x^{k-1}dx$$

are evaluated to be $\Gamma(k)/\lambda^k$ simply by substitution. For example, the nth moment of the gamma distribution

$$\mu_n=\int_0^\infty x^n\lambda^k e^{-\lambda x}x^{k-1}/\Gamma(k)dx$$

$$=\frac{\lambda^k\Gamma(n+k)}{\Gamma(k)\lambda^{n+k}}$$

$$=(k)_n\lambda^{-n}$$

[†]Compare Problem 55, Section 2.9.

is easily found without performing the integration again. Similarly, the variance of this distribution can be shown to be k/λ^2 and the Laplace transform to be $[\lambda/(s+\lambda)]^k$.

4.3. Change of Variable

In the study of discrete random variables, the change from the distribution of one random variable to another functionally related one was basically substitution, with the probabilities unchanged. A little more care must be taken in the continuous cases, although the result is rather similar. Beginning with the basic relationship between the random variable and its probabilities

$$P(X<x)=F(x),$$

suppose $Y=h(X)$, where h is a strictly monotonic function with a continuous derivative.[†] Then

$$P(Y<x)=P\big(h(X)<x\big)$$

$$=P\big(X<h^{-1}(x)\big).$$

In the continuous case, the density function $f_1(x)$ of Y and $f_2(x)$ of X are thus given by

$$f_1(x)=\frac{d}{dx}F\big(h^{-1}(x)\big)$$

$$=f_2\big(h^{-1}(x)\big)\frac{d}{dx}\big(h^{-1}(x)\big). \tag{1}$$

This simply says that the substitution must be made not only in the density function, but also in the differential element. Thus, for example, if X is exponential with parameter λ, the distribution of kX is not $\lambda e^{-\lambda y/k}$, but rather $(\lambda/k)e^{-\lambda y/k}$, as normalization would show in any case. The extra factor $1/k$ can be thought of as coming from the substitution of dy/k for dx.

[†] The student with experience in analysis will see why these conditions are necessary. In virtually all cases of importance in this book, the relationship is linear.

Similarity (and intuitively), when more than one random variable is involved in a joint distribution, the substitution of one joint density for another must involve the Jacobian, for example,

$$dx\,dy = \begin{vmatrix} \dfrac{\partial x}{\partial u} & \dfrac{\partial x}{\partial v} \\[2mm] \dfrac{\partial y}{\partial u} & \dfrac{\partial y}{\partial v} \end{vmatrix} du\,dv = J\,du\,dv. \tag{2}$$

Example 1. If X and Y are jointly and independently exponential, with parameters λ and μ, respectively, then their joint density function is

$$\lambda\mu e^{-\lambda x - \mu y}.$$

Now, let $U = X + Y$, $V = X/Y$, each defined over $(0, \infty)$. Then

$$J = \frac{u}{(1+v)^2}$$

and the joint density function of U and V is

$$\frac{\lambda\mu u}{(1+v)^2} \exp\left(-\frac{\lambda v + \mu}{1+v} u \right),$$

and the single-variable densities of U and V can be found by integrating out the other variable. In cases where the joint density factors into separate densities (not in the example just given), the variables are independent. For example, the student should verify that if X and Y are independently exponential, with the same parameter, then U and V are also independent, the former being gamma distributed and the latter being distributed with density $(1+v)^{-2}$.

When random variables are mixed discrete and continuous, their distributions are not characterized by a simple probability density function, and therefore the method given in this section does not apply, except that the substitution in the distribution function $F(x)$ remains valid.

4.4. Convolutions of Density Functions

Just as with discrete random variables, it is frequently desirable to find the distribution of the sums of independent, continuous random variables. Suppose X, Y, and $X + Y$ have respectively density functions $f(x)$, $g(x)$, and

$h(x)$, with cumulative forms

$$F(x)=\int_0^x f(u)du, \qquad G(x)=\int_0^x g(u)du, \qquad H(x)=\int_0^x h(u)du,$$

then

$$H(x)=P(X+Y<x)=\int\int_{u+v<x} f(u)g(v)du\,dv$$

$$=\int_0^\infty \int_0^{x-v} f(u)g(v)du\,dv$$

$$=\int_0^\infty F(x-v)g(v)dv. \qquad (3)$$

Differentiating with respect to x gives the convolution relationship for density functions of wholly continuous random variables:

$$h(x)=f(x)*g(x)=\int_0^x f(x-v)g(v)dv. \qquad (4)$$

The upper limit is now the finite value x, since the density functions under consideration are zero for negative arguments.

It will be noted that the convolution formula given is entirely analogous to the formula for discrete random variables as given in Section 2.6, and, like the earlier formula, the functions f and g are interchangeable.

Also, for students interested in the theory of integration, it is clear that necessary conditions for differentiation under the integral sign are fulfilled in the present instance.

Actually working out a convolution can be tricky in the continuous case because of the range of the variable. Consider the following example. Let X be a random variable with a continuous rectangular distribution over $(0, a)$,

$$f(x)=\frac{1}{a}, \qquad 0<x<a,$$

and let Y be a negative exponential random variable,

$$g(x)=\mu e^{-\mu x}, \qquad 0<x<\infty.$$

What is the distribution of $X+Y$, under the assumption of independence? If Eq. (4) is applied quite mechanically, one might suppose the answer to be

$$\int_0^x \frac{1}{a}\mu e^{-\mu x}dx = \frac{1}{a}(1-e^{-\mu x}), \qquad 0<x<\infty,$$

which cannot be correct, since the right side does not even integrate to unity. The important fact which is omitted in this calculation is that the rectangular distribution vanishes outside the range $(0, a)$, so that the integrand can also vanish for certain ranges of the variable. In fact, although it is true that $X+Y$ has range $(0, \infty)$, the evaluation of the convolution integral requires separate consideration of the ranges $(0, a)$ and (a, ∞). In the first case, the upper limit x is correct, but if $x>a$, the upper limit must be replaced by a since the rectangular distribution is zero for values larger than a. Thus the density for $X+Y$ is

$$\int_0^a \frac{1}{a}\mu e^{-\mu(x-t)}dt = \frac{1}{a}e^{-\mu x}(e^{\mu a}-1), \qquad x>a,$$

$$\int_0^x \frac{1}{a}\mu e^{-\mu(x-t)}dt = \frac{1}{a}(1-e^{-\mu x}), \qquad x<a.$$

It is left as an exercise for the student to show that the gamma distribution with parameters k and λ is the k-fold convolution of the exponential distribution with parameter λ.

As in the discrete case, starred exponents will be used to denote repeated convolutions of a distribution with itself:

$$[f(x)]^{n*}=f(x)*f(x)* \cdots *f(x) \qquad (n \text{ "factors"}).$$

4.5. The Incomplete Gamma Function

In Section 1.9, the gamma function was defined by the integral

$$\Gamma(n)=\int_0^\infty e^{-t}t^{n-1}dt, \qquad n>0.$$

If the range of integration is split in two parts by an arbitrary value x, $0<x<\infty$, the separate integrals are called *incomplete gamma functions* and are denoted by

$$\gamma(n, x) = \int_0^x e^{-t} t^{n-1} dt, \tag{5}$$

the *lower* incomplete gamma function, and by

$$\Gamma(n, x) = \int_x^\infty e^{-t} t^{n-1} dt, \tag{6}$$

the *upper* incomplete gamma function. Naturally

$$\gamma(n, x) + \Gamma(n, x) = \Gamma(n).$$

These functions are useful in many ways; in probability their chief roles are as the distribution functions for two important distributions: the gamma distribution and the Poisson distribution. This link between two prominent distributions, one continuous and the other discrete, turns out to be especially significant in studying stochastic point processes, beginning in Section 5.3.

The Gamma Distribution (with parameters λ and k)

$$F(x) = \int_0^x \frac{e^{-\lambda t} t^{k-1} \lambda^k}{\Gamma(k)} dt = \frac{\gamma(k, \lambda x)}{\Gamma(k)}, \tag{7}$$

so that the tail of the distribution (upper integral) is $\Gamma(k, \lambda x)/\Gamma(k)$.

It is left as an exercise (Problem 46) to show that the difference of incomplete gamma functions is exactly a Poisson term:

$$\frac{\gamma(n, x)}{\Gamma(n)} - \frac{\gamma(n+1, x)}{\Gamma(n+1)} = \frac{x^n e^{-x}}{n!} \tag{8}$$

The Poisson Distribution (with parameter λ). The distribution function $P(x) = P(X < x)$ (see Section 1.8) is defined by

$$P(x) = \sum_{j=0}^{x-1} \frac{\lambda^j e^{-\lambda}}{j!},$$

which can be written, using Eq. (8),

$$P(x)=e^{-\lambda}+\sum_{j=1}^{x-1}\left(\frac{\gamma(j,\lambda)}{\Gamma(j)}-\frac{\gamma(j+1,\lambda)}{\Gamma(j+1)}\right)$$

$$=1-\frac{\gamma(x,\lambda)}{\Gamma(x)}$$

$$=\frac{\Gamma(x,\lambda)}{\Gamma(x)}. \tag{9}$$

Similarly, the tail $P(X \geq x)$ of the Poisson distribution can be written

$$\sum_{j=x}^{\infty}\frac{e^{-\lambda}\lambda^{j}}{j!}=\frac{\gamma(x,\lambda)}{\Gamma(x)}. \tag{10}$$

This curious, nearly identical, relationship between the upper part of the gamma distribution and the lower part of the Poisson distribution (and vice versa) reflects an interesting probabilistic fact.

Consider a sequence of points on a line defined by the property that the distribution of the distances between neighboring points is independently exponential. Specifically, if the points have coordinates $\xi_0 = 0$, $\xi_1, \xi_2, \ldots$, and if

$$P(\xi_{n+1}-\xi_n<x)=1-e^{-\lambda x}, \qquad n=0,1,2,\ldots,$$

then, since the n-fold convolution of the negative exponential distribution is the gamma distribution,

$$P(\xi_{n+k}-\xi_n<x)=\frac{\gamma(k,\lambda x)}{\Gamma(k)}, \qquad k=1,2,3,\ldots, \; n=0,1,2,\ldots,$$

this being the distribution function for the length of a line segment joining two of the points which are separated by $k-1$ other points. The density function for this distance is thus a gamma distribution function with parameters k and λ; in the special case $k=1$, it is exponential. Now consider a segment of the line of length L which does not contain the origin, and let Y denote the number of points lying on L. Then, for example, the leftmost point (if any) lying within L is ξ_n,

$$P\left(Y \overset{\geq}{=} k\right) = P(\xi_{n+k} - \xi_n < L)$$

$$= \frac{\gamma(k, \lambda L)}{\Gamma(k)} \qquad \left[\text{by Eq. (7)}\right]$$

$$= \sum_{j=k}^{\infty} \frac{e^{-\lambda L}(\lambda L)^j}{j!} \qquad \left[\text{by Eq. (10)}\right].$$

This shows that if the *gaps* between points are exponentially distributed with parameter λ, then the *counts* of points in length L are Poisson distributed with parameter λL. The argument, which will be generalized in discussions of renewal processes, can also be reversed, giving the *Poisson–gamma* relation. Naturally, the mean count in a unit interval is λ, whereas the mean gap is $1/\lambda$.

An important application of this relationship occurs when the line represents a time axis and the points are events in time. If a distribution is formed by counting events in time, such as cars passing a point per minute, particles arriving at a particle counter per minute, mine disasters per year, fire calls per day, or wars per century, then this distribution is Poisson if and only if the continuous distribution formed by the time between cars, particles, disasters, fire calls, or wars is independent and exponential. Such events are sometimes called *random*, or Poisson. Because of the more general use of the word "random" to mean "not deterministic," this volume uses "Poisson" to indicate such a point sequence. It becomes important, then, to distinguish clearly between the two meanings: A Poisson sequence has a Poisson *counting distribution*.

Two other facts which have already been established shed additional light on Poisson events. The Poisson distribution counts points placed independently on a segment (Section 2.4); and if one point follows another with exponential distribution, then after a fixed period of time, the distribution of the gap to the next point is still exponential and with the same parameter (Section 4.2). This characterization is referred to by saying that the exponential distribution is *memoryless*.

4.6. The Beta Distribution and the Incomplete Beta Function

A continuous distribution over the finite interval $(0, 1)$ can be formed from the function

$$Cx^{p-1}(1-x)^{q-1}, \qquad p>0, q>0.$$

It is shown in Section 1.9, Eq. (18), that the normalizing constant must be

$$C = \frac{\Gamma(p+q)}{\Gamma(p)\Gamma(q)}.$$

This two-parameter distribution can assume a wide variety of shapes—including the continuous rectangular—depending on the values of the parameters. It can also, by changes in parameters and range, be made to cover an arbitrary, finite interval (a, b). It is not difficult to show that the mean is $p/(p+q)$ and that the variance is $pq/[(p+q)^2(p+q+1)]$.

The *incomplete beta function* is defined similarly to the incomplete gamma function, by the integral

$$B_x(p, q) = \int_0^x t^{p-1}(1-t)^{q-1} dt. \tag{11}$$

Also similarly to the incomplete gamma function, it can be used to represent the distribution function of two probability distributions, the one continuous and the other discrete.

The Beta Distribution. It is clear from the definitions that, for the beta distribution,

$$F(x) = \frac{B_x(p, q)}{B(p, q)}. \tag{12}$$

The Binomial Distribution. The distribution function can be represented as follows:

$$1 - P(x) = \frac{B_\pi(x, n-x+1)}{B(x, n-x+1)}. \tag{13}$$

The proof of Eq. (13) is not difficult, but it can be tedious. It consists of the repeated integration by parts of the definition of the incomplete beta function; the process ejects successive binomial terms. The first step, for example, yields

$$B_\pi(x, n-x+1) = \frac{1}{x}(1-\pi)^{n-x}\pi^x + \frac{n-x}{x}B_\pi(x+1, n-x).$$

When divided by the entire beta function $B(x, n-x+1) = (x-1)!(n-x)!/n!$, the first term becomes the xth binomial term. Similarly, further integration by parts produces the $(x+1)$st, $(x+2)$nd,... binomial terms, while at the same time increasing the first beta argument and decreasing the second one. The final beta term,

$$\frac{(n-x)(n-x-1)\cdots 3\cdot 2\cdot 1}{x(x+1)\cdots(n-1)}B_\pi(n, 1),$$

easily turns into the nth binomial term, thus showing that the ratio of the incomplete to the complete beta functions is the tail $P(X \geq x)$ of the binomial distribution.

This *beta–binomial* relationship is almost exactly parallel to the *Poisson–gamma* relationship of Section 4.5. A probabilistic interpretation is as follows. Suppose numbers $X_1, X_2,..., X_n$ are chosen independently, at random, on the unit interval. Let $Y_1 \leq Y_2 < \cdots \leq Y_n$ be the same numbers designated in order of magnitude. Let X be the number of the X_j which lie in the interval $(0, \pi)$, so that X is a binomial random variable with parameters n and π. Then the event $X \geq x$ is the same as the event $Y_x \leq \pi$. The first inequality has probability equal to the left side of Eq. (13), and therefore the beta distribution on the right side gives the probability of the second inequality. Y_j is called in statistics an *ordered sample*, a concept of considerable importance.

4.7. Parameter Mixing

Consider the binomial–Poisson limit (Section 2.4), with the interpretation that the number of events in time t is Poisson distributed with parameter μt, under the assumption of independence of the events. According to one model for accident occurrence, this adequately describes the probability of X accidents in time t; according to a refinement of the same theory, μ itself may be a random variable so that the mean number of accidents in a given time varies in the population.[†] Then the Poisson distribution would play the role of a conditional distribution, conditional on a fixed value of μ, and an unconditional distribution would be obtained by using a continuous version of Eq. (6) of Section 2.3, namely,

$$P(X=x)=\int_0^\infty \frac{e^{-\mu t}(\mu t)^x}{x!} f(\mu)d\mu, \qquad (14)$$

where $f(\mu)$ is the density function for the random variable μ. Note that the range of μ is $(0,\infty)$, so that it would be natural to choose a density function $f(\mu)$ defined over this range. One choice that seems logical, and that has in fact been widely used, is the gamma distribution, say, with parameters λ and k. Then, using the substitution technique given in Section 4.2 in connection with the gamma mean, the integral is evaluated as

$$P(X \doteq x) = \frac{t^x \lambda^k \Gamma(x+k)}{x! \Gamma(k)(t+\lambda)^{x+k}}$$

$$= \binom{x+k-1}{x}\left(\frac{\lambda}{\lambda+t}\right)^k \left(\frac{t}{\lambda+t}\right)^x,$$

a negative binomial distribution. This model, when used in accident analysis, is often referred to as an "accident proneness" model, with the parameter μ called the proneness. In the theory of probability, the Poisson distribution is said to be *mixed* with the gamma distribution to give the negative binomial. This use of the term "mixed" should not be confused with the same word as applied to a distribution which is partly discrete and partly continuous[‡]; in case of doubt, it is better to write "parameter mixed" in one case and "discrete and continuous mixed" in the other.

There are many other examples of mixing which are of some importance. The beta distribution, being defined over the unit interval, is useful as a mixing distribution when the parameter in question is a probability, such as ρ in the geometric distribution or π in the binomial.

[†]The simpler interpretation is due to L. von Bortkewitsch (Russian/German statistician, 1868–1931) in his monograph, *Das Gesetz der kleinen Zahlen*, B. G. Teubner, Leipzig. The more complex view (now called "accident proneness") was developed by M. Greenwood and G. U. Yule in their 1920 paper, An inquiry into the nature of frequency distributions representative of multiple happenings, with particular reference to the occurrence of multiple attacks of disease or of repeated accidents, *J. Roy. Stat. Soc.* **83**, 255–279.

[‡]However, an artifice can be constructed which would give a mixed discrete and continuous distribution as a special case of parameter mixing.

The Geometric Mixed Distribution and Beta Mixing

$$P(X=x)=\int_0^1(1-\rho)\rho^x\frac{\Gamma(p+q)}{\Gamma(p)\Gamma(q)}\rho^{p-1}(1-\rho)^{q-1}d\rho$$

$$=\frac{\Gamma(p+q)\Gamma(x+p)\Gamma(1+q)}{\Gamma(p)\Gamma(q)\Gamma(x+p+q+1)}$$

$$=\frac{B(x+p,1+q)}{B(p,q)},\qquad x=0,1,2,\dots. \tag{15}$$

A general model leading to this distribution could be stated in the following terms. If the probability is ρ that a trial will result in a success, then the geometric distribution gives the probability of a run of x successes followed by a failure. In an application, a success might represent an item in quality control which passes inspection. If the manufacturing process permitted ρ to vary according to a beta distribution, the probability of a sequence of x items passing inspection would be given by the distribution (15), where the parameters are now p and q, determined by the manufacturing process, rather than ρ, determined by the inspection process.

The Binomial Mixed Distribution and Beta Mixing

$$P(X=x)=\int_0^1\binom{n}{x}\pi^x(1-\pi)^{n-x}\frac{\Gamma(p+q)}{\Gamma(p)\Gamma(q)}\pi^{p-1}(1-\pi)^{q-1}d\pi$$

$$=\frac{\binom{n}{x}B(x+p,n+q-x)}{B(p,q)},\qquad x=0,1,\dots,n. \tag{16}$$

This distribution is called the *negative hypergeometric*; it has been used to model the probability of X chromosome associations, where π is the probability of an association, which varies from nucleus to nucleus.

It is also possible to mix with respect to a discrete distribution when the parameter assumes discrete values. This is the case with the other binomial parameter.

The Binomial Mixed Distribution and Poisson Mixing

$$P(X=x)=\sum_{n=0}^{\infty}\binom{n}{x}\pi^x(1-\pi)^{n-x}\frac{e^{-\lambda}\lambda^n}{n!}.$$

It is left as an exercise for the student to show that in this case X is unconditionally Poisson distributed with parameter $\lambda\pi$ and to construct an example which this situation might reasonably be expected to model.

From a theoretical point of view, parameter mixing is equivalent to the selection of a single random variable from a sequence of random variables. This interpretation can, in concrete circumstances, appear paradoxical. Thus, in the accident proneness example, given a population of individuals, each of whom experiences a Poisson-distributed number of accidents in a certain time, the selection of an individual from the population would seem to result in a person having negative binomially distributed accidents. However, this is not quite the implication of the model. To obtain the negative binomial distribution, it would be necessary to choose different individuals according to the same probabilistic scheme for each time period. Under this circumstance, the distribution built up over many time periods would be negative binomial, although none of the individuals concerned would have a negative binomial accident experience.

Thus, if parameter mixing is used as the justification for a negative binomial accident distribution, the population at risk cannot be always the same, but must be randomly chosen afresh for each time period.

4.8. Distribution Functions

In Section 4.2, it was shown that continuous and discrete random variables could be defined in the same mathematical format by using the distribution function rather than the discrete probabilities or the continuous probability density function. It is also true that mixed (discrete and continuous) random variables can be described in this way: For the wholly continuous case, the distribution function is continuous; for the discrete case, it is a step function, constant between jumps, and for the mixed case, it is partly continuous and partly a step function.

Formally, a real-valued function $F(x)$ is called a distribution function if it satisfies the following conditions:

1. It is defined over $(-\infty, \infty)$, with values $F(-\infty)=0$, $F(\infty)=1$. In the present volume, the additional assumption will be made that $F(x)=0$ for $x\leq 0$, thus confining the discussion to positive random variables.
2. It is nondecreasing, that is, $x<y$ implies $F(x)\leq F(y)$.
3. It is continuous on the left, that is, $\lim_{y\uparrow x} F(y)=F(x)$.

The third assumption means that the functional value at points of discontinuity are associated with the lower branch of the curve, since $F(x)=P(X<x)$.[†]

Special Cases. (a) If $F(x)$ is continuous, and $P(X<x)=F(x)$, then X is a *continuous* random variable.

(b) If $F(x)$ is a step function, and $P(X<x)=F(x)$, then X is a *discrete* random variable.

(c) If $F(x)$ is continuous except at a countable number of points, where $F(x)$ has jump discontinuities, and if there is an interval in which $F'(x)$ exists and is nonzero, and $P(X<x)=F(x)$, then X is a *mixed* (discrete and continuous) random variable.

If $F'(x)$ exists and is continuous for all $x\geq0$, then a familiar theorem from calculus guarantees that for all $x>0$,

$$F(x)=\int_0^x F'(t)\,dt. \tag{17}$$

But if X is a discrete or mixed random variable, $F'(x)$ does not even exist at the points where F is discontinuous.[‡]

For a discrete distribution, let the jumps occur at the points a_x, and let the magnitude of the jump at a_x be p_x. Define the *delta function* $\Delta(x)$ by

$$\Delta(x)=0, \qquad x\leq0,$$

$$\Delta(x)=1, \qquad x>0.$$

Thus $\Delta(x-t)$ is the distribution function for the causal distribution at the value t. Then, for any discrete distribution, the distribution function can be written

$$F(x)=\sum_j p_j\Delta(x-a_j). \tag{18}$$

[†]As pointed out in Section 1.8, an equivalent theory can be developed on the basis of right continuity, that is, with $F(x)=P(X\leq x)$.

[‡]By using Lebesgue integration, a version of Eq. (17) can be obtained even if $F'(t)$ is not defined at all points; however, the resulting theorem is not useful here because it holds only if F is continuous. Indeed F must satisfy a stronger condition called absolute continuity. See, for example, ROYDEN, H. L. (1968), *Real Analysis*, Macmillan, New York.

In the third case, mixed distributions, it is necessary to separate $F_1(x)$, represented by Eq. (17), and $F_2(x)$, represented by Eq. (18); then

$$F(x) = \pi F_1(x) + (1-\pi)F_2(x), \tag{19}$$

where π is the "share" of the continuous part of the distribution function.

For example, in a speed distribution, if one-fourth of vehicles are stationary, with the remainder having speeds distributed exponentially with parameter λ, then

$$F(x) = 0, \qquad x \leq 0,$$

$$F(x) = \tfrac{1}{4} + \tfrac{3}{4}\left(1 - e^{-\lambda x}\right), \qquad x > 0.$$

Here $\pi = \tfrac{3}{4}$, $F_2(x) = \Delta(x) = 1$.

Combining the three types of probability distributions in a single format is not simply a mathematician's exercise—it is important in studying mixed distributions. When a random variable is partly discrete and partly continuous, familiar calculations such as those leading to moments and generating functions, convolutions, and so forth, would appear to be partly summations and partly integrations. It is necessary, therefore, to develop a mathematical technique which will unify summation and integration; this technique is called Stieltjes integration.

4.9. Stieltjes[†] Integration

The ordinary ("Riemann") integral is constructed in a sequence of steps: (i) A *function* is defined over a finite interval and is bounded, (ii) the interval is divided into subintervals, (iii) the length of each subinterval is multiplied by a functional value in the subinterval and the results added together, and (iv) the limit is taken as the subdivision becomes finer in a way which insures that the limit of the length of each subinterval is zero. When such a limit exists, and is independent of the various intermediate choices, it is called the (Riemann) integral of the function over the interval. These familiar integrals have many well-known properties, such as integration by parts and the indefinite integral being an inverse derivative. Riemann integrals are, however, limited in certain ways, one of which is in

[†] Thomas Jan Stieltjes, Dutch mathematician, 1856–1894.

dealing with discontinuous functions, such as a distribution function for a discrete or mixed distribution.

It is convenient, therefore, to consider a generalization of the Riemann integral, the Stieltjes integral. In this section an outline is given of some of the more important steps in the definition of Stieltjes integration.

Consider functions $g(x)$ and $F(x)$, both of which are defined for the interval $a \leq x \leq b$. Let S denote a subdivision of the interval by points $x_0, x_1, \ldots, x_n$, where

$$a = x_0 < x_1 < \cdots < x_n = b.$$

Let

$$\delta = \text{norm}(S) = \max_i (x_{i+1} - x_i),$$

and let $\xi_0, \xi_1, \ldots, \xi_{n-1}$ be chosen so that $x_i \leq \xi_i \leq x_{i+1}$. Then, if the limit

$$\lim_{\delta \to 0} \sum_{i=0}^{n-1} g(\xi_i)[F(x_{i+1}) - F(x_i)] \tag{20}$$

exists independently of the choice of subdivision and of the selection of points ξ_i, its value is called the *Stieltjes integral* of $g(x)$ with respect to $F(x)$ over the interval (a, b) and it is written

$$\int_a^b g(x) dF(x). \tag{21}$$

It will be noted that the Stieltjes integral becomes a Riemann integral when $F(x) = x$, according to the definition, and also that the notation chosen for Eq. (21) reflects this fact. Just as with the Riemann integral, one of the first tasks is to establish conditions under which the Stieltjes integral exists. Since the application to probability (Laplace transforms of distribution functions) is rather special, no discussion will be given of general necessary conditions, but only of those sufficient for this purpose.

Theorem 1. If, on the interval $[a, b]$, $g(x)$ is continuous and $F(x)$ is monotonic nondecreasing, then the integral (21) exists.

Proof. For a given subdivision $\mathcal{S}$ let

$$\left.\begin{array}{l} M_i = \max g(x), \\ m_i = \min g(x), \end{array}\right\} \quad x_i \le x \le x_{i+1}, i = 0, 1, \ldots, n-1,$$

so that $m_i \le g(\xi_i) \le M_i$. Since $F(x)$ is monotonic nondecreasing, $\Delta F_i = F(x_{i+1}) - F(x_i) \ge 0$. Therefore

$$\sum_{i=0}^{n-1} m_i \Delta F_i \le \sum_{i=0}^{n-1} g(\xi_i) \Delta F_i \le \sum_{i=0}^{n-1} M_i \Delta F_i.$$

The left and right terms of the inequality are called the *lower* and *upper* *sums*, respectively, of the subdivision $\mathcal{S}$, and are written $L(\mathcal{S})$ and $U(\mathcal{S})$. If $\mathcal{S}$ and $\mathcal{S}'$ are any two subdivisions of the interval, $L(\mathcal{S}) \stackrel{<}{=} U(\mathcal{S}')$; the proof of this fact is left as an exercise for the student and is based on considering a subdivision $\mathcal{S}''$ consisting of all points in either $\mathcal{S}$ or $\mathcal{S}'$. Now define two numbers A and B by the equations

$$A = \sup \sum_{i=0}^{n-1} m_i \Delta F_i,$$

$$B = \inf \sum_{i=0}^{n-1} M_i \Delta F_i,$$

so that $A \le B$, and thus

$$L(\mathcal{S}) \le A \le B \le U(\mathcal{S}).$$

To prove that the integral exists, it is sufficient to show that

$$\lim_{\delta \to 0} \left[U(\mathcal{S}) - L(\mathcal{S}) \right] = 0,$$

that is, given $\varepsilon > 0$, the norm δ can be chosen so that

$$U(\mathcal{S}) - L(\mathcal{S}) < \varepsilon.$$

Since $g(x)$ is assumed to be continuous in the interval $[a, b]$, it is possible to find a constant δ' such that $|x'-x''|<\delta'$ implies

$$g(x')-g(x'')<\frac{\varepsilon}{F(b)-F(a)}.$$

Let the norm δ be any value less than this δ': $\delta<\delta'$, and take for x' and x'' successively the values x_i, x_{i+1}, defining $\mathcal{S}$. Then

$$U(\mathcal{S})-L(\mathcal{S})=\sum_{i=0}^{n-1}(M_i-m_i)\Delta F_i$$

$$<\sum_{i=0}^{n-1}\frac{\varepsilon}{F(b)-F(a)}\Delta F_i$$

$$<\frac{\varepsilon}{F(b)-F(a)}\sum_{i=0}^{n-1}\Delta F_i$$

$$<\varepsilon. \qquad\qquad\qquad \square$$

Theorem 2. If $g(x)$ is integrable with respect to $F(x)$ on $[a, b]$, then $F(x)$ is integrable with respect to $g(x)$ on $[a, b]$, and

$$\int_a^b g(x)dF(x)+\int_a^b F(x)dg(x)=F(b)g(b)-F(a)g(a). \qquad (22)$$

Note. This theorem gives the integration by parts formula familiar both in Riemann and in Stieltjes integration.

Proof. The sum $\sum_{i=0}^{n-1}g(\xi_i)\Delta F_i$ can be rearranged in the form

$$\sum_{k=1}^{n}F(x_k)[g(\xi_{k-1})-g(\xi_k)]+F(b)g(b)-F(a)g(a).$$

As the maximum of the differences $x_{i+1}-x_i$ approaches zero, so will the maximum of the differences ξ_0-a, $\xi_1-\xi_0,\ldots,\xi_{n-1}-\xi_{n-2}$, $b-\xi_{n-1}$. Therefore the rewritten sum approaches the required integral. $\square$

Many other familiar properties of Riemann integrals are also properties of Stieltjes integrals, and the proofs follow familiar lines and need not be repeated. For reference, some of the more important relations are stated.

Theorem 3

$$\int_a^b g(x)dF(x)=\int_a^c g(x)dF(x)+\int_c^b g(x)dF(x).$$

Theorem 4

$$\int_a^b [g_1(x)+g_2(x)]dF(x)=\int_a^b g_1(x)dF(x)+\int_a^b g_2(x)dF(x).$$

Theorem 5

$$\int_a^b g(x)d[F_1(x)+F_2(x)]=\int_a^b g(x)dF_1(x)+\int_a^b g(x)dF_2(x).$$

Theorem 6

$$\int_a^b cg(x)dF(x)=\int_a^b g(x)dF(cx)=c\int_a^b g(x)dF(x).$$

Improper integrals (i.e., those with upper limit infinity) are important and are defined, as with Riemann integrals, as limits.

Definition 1

$$\int_a^\infty g(x)dF(x)= \lim_{R\to\infty} \int_a^R g(x)dF(x).$$

In some cases the Stieltjes integral can be reduced at once to a Riemann integral.

Theorem 7. If $F(x)$ is differentiable in $[a, b]$, with $F'(x)=f(x)$, then

$$\int_a^b g(x)dF(x)=\int_a^b g(x)f(x)dx. \tag{23}$$

Theorems 2 and 7 provide means for evaluating many Stieltjes integrals—Theorem 7 when F is differentiable, and Theorem 2 when g is differentiable. It should be emphasized, however, that this by no means permits an easy evaluation of all Stieltjes integrals; indeed, in some quite simple cases (such as that where g and F are discontinuous at the same point) the integral does not exist.

As an example of the two methods, consider $\int_{-1}^{+1} e^x d|x|$. By one method,

$$\int_{-1}^{+1} e^x d|x| = \int_{-1}^{0} e^x d|x| + \int_{0}^{+1} e^x d|x|$$

$$= \int_{-1}^{0} e^x d(-x) + \int_{0}^{+1} e^x dx$$

$$= -\int_{-1}^{0} e^x dx + \int_{0}^{+1} e^x dx$$

$$= e + e^{-1} - 2,$$

while, by Theorem 2, the calculation is simpler:

$$\int_{-1}^{+1} e^x d|x| = -\int_{-1}^{+1} |x| de^x + e + e^{-1}$$

$$= -\int_{-1}^{0} (-x)e^x dx - \int_{0}^{1} xe^x dx + e + e^{-1},$$

which gives the same result.

Rather than pursue Stieltjes integration in general, it will be more useful to consider it in the probability context of finding integrals with respect to distribution functions. Note: It is in the development of Stieltjes integration that the importance of left continuity for distribution functions becomes essential. For, using the definition of Stieltjes integration, it follows immediately that

$$\int_{a}^{b} g(x) dF(x) = 0,$$

whenever $F(x)$ is constant over the closed interval $[a, b]$. Suppose $g(x) = 1$ and consider the deterministic distribution at the origin with distribution function $F(x)$. With left continuity, $F(x)$ is not constant over the closed interval $[0, b]$ and so this formula does not apply. But with right continuity, $F(x) = 1$ over the closed interval $[0, b]$, and so the formula would require that the total probability of the deterministic distribution would be zero, rather than unity, as needed for probability theory. The only way out of the difficulty would be to significantly complicate the formulation by substituting limits from the left in place of fixed lower limits of integration, i.e., to write always

$$\lim_{x \uparrow a} \int_x^b$$

in place of

$$\int_a^b .$$

It is easier and more pleasant to use left continuity from the beginning.

4.10. The Laplace Transform

Let $F(x)$ be a (continuous, discrete, or mixed) distribution function defined over $(0, \infty)$, with $F(x)=0$, $x \leq 0$. The Laplace–Stieltjes transform $\phi(s)$ of $F(x)$ is defined to be

$$\phi(s) = \int_0^\infty e^{-sx} dF(x), \tag{24}$$

provided the integral converges.

Since e^{-sx} is continuous and $F(x)$ is monotonic, the integral over any finite interval must, by Theorem 1 of Section 4.9, converge. This does not, however, guarantee that the improper integral (24) is finite.

Theorem 1. If the Laplace–Stieltjes transform $\phi(s)$ of $F(x)$ converges, then

$$\int_0^\infty e^{-sx} dF(x) = s \int_0^\infty e^{-sx} F(x) dx.$$

Proof. This follows directly from Theorem 2 of Section 4.9, since $F(0)=0$, $F(\infty)=1$, $e^{-\infty}=0$. The integral on the right side of this equation is called the Laplace transform of $F(x)$. $\qquad \square$

Theorem 2. If X is a continuous random variable defined over $(0, \infty)$, with $F(x) = P(X < x)$ and $F'(x) = f(x)$, then

$$\int_0^\infty e^{-sx} dF(x) = \int_0^\infty e^{-sx} f(x) dx.$$

Proof. This follows from Theorem 7 of Section 4.9. $\qquad\qquad\square$

Theorem 3. If X is a discrete random variable with non-negative integral values $p_x = P(X = x)$, $x = 0, 1, 2, \ldots$ and with probability generating function $\beta(s) = \Sigma s^x p_x$, then for $n < x \le n+1$, $F(x) = p_0 + p_1 + \cdots + p_n$, and

$$\int_0^\infty e^{-sx} dF(x) = \beta(e^{-s}).$$

This means that for discrete random variables, the Laplace–Stieltjes transform is only the probability generating function with s replaced by e^{-s}.

Proof. It is necessary to integrate over the separate steps of the step function, keeping in mind that $F(x) = F(n+1)$ for $n < x \le n+1$:

$$\int_0^\infty e^{-sx} dF(x) = s \int_0^\infty e^{-sx} F(x) dx$$

$$= \sum_{n=0}^\infty s \int_n^{n+1} e^{-sx} F(x) dx$$

$$= \sum_{n=0}^\infty s \int_n^{n+1} e^{-sx} F(n+1) dx$$

$$= \sum_{n=0}^\infty (e^{-sn} - e^{-s(n+1)}) F(n+1)$$

$$= [F(1) - e^{-s} F(1)] + [e^{-s} F(2) - e^{-2s} F(2)] + \cdots$$

$$= p_0 + e^{-s} p_1 + e^{-2s} p_2 + \cdots$$

$$= \beta(e^{-s}).\qquad\qquad\square$$

A similar theorem could be given for general mixed discrete and continuous distributions, but it will be more useful to consider certain special cases.

Example 1. A distribution consists of a discrete element $P(X=0)=\alpha$ and a continuous element $P(X<x)=(1-\alpha)(1-e^{-\lambda x})$, that is, a probability α of a zero value, and the remainder distributed exponentially with parameter λ, and total probability $1-\alpha$. Then

$$F(x)=\alpha+(1-\alpha)(1-e^{-\lambda x}), \qquad x>0$$

$$\phi(s)=\int_0^\infty e^{-sx}dF(x)$$

$$=s\int_0^\infty e^{-sx}F(x)dx$$

$$=s\int_0^\infty e^{-sx}\left[\alpha+(1-\alpha)(1-e^{-\lambda x})\right]dx$$

$$=\alpha+(1-\alpha)s\int_0^\infty e^{-sx}(1-e^{-\lambda x})dx$$

$$=\alpha+(1-\alpha)-(1-\alpha)s\int_0^\infty e^{-(\lambda+s)x}dx$$

$$=1-(1-\alpha)\frac{s}{s+\lambda}$$

$$=\frac{\alpha s+\lambda}{s+\lambda}.$$

This calculation may suggest, and is at least consistent with, another way of finding Laplace–Stieltjes transforms in the mixed case, namely to combine Theorems 2 and 3, using Theorem 2 for the continuous portion and Theorem 3 for the discrete portion. In the example, the discrete probability generating function for an element of probability α at the origin is simply $\beta(s)=\alpha$. The Laplace transform of the density $\lambda e^{-\lambda x}$ of the exponential distribution is $\lambda/(\lambda+s)$, and this has probability $1-\alpha$. Hence

$$\int_0^\infty e^{-sx}dF(x)=\alpha+(1-\alpha)\frac{\lambda}{\lambda+s}$$

$$=\frac{\lambda+\alpha s}{\lambda+s},$$

as before. Calculations of this nature can often be organized algebraically by using the function $\delta(x)$, which represents the causal distribution at the origin [the corresponding distribution function was given in Section 4.8 as $\Delta(x)$],

$$\delta(x)=1, \qquad x=0,$$

$$\delta(x)=0, \qquad x\neq0,$$

in the following way:

$$\int_0^\infty e^{-sx}dF(x)=\alpha\int_0^\infty e^{-sx}\delta(x)dx+(1-\alpha)\int_0^\infty e^{-sx}\lambda e^{-\lambda x}dx,$$

with the understanding that when $\delta(x-d)$ occurs in an integrand, it is a sign that the integral is obtained simply by setting $x=d$ in the remainder of the integrand.

This is a handy procedure which can be justified or made mathematically rigorous by using more advanced theory in analysis, i.e., the theory of generalized functions developed by the French mathematician L. Schwartz.[†] For present purposes it must be considered to be only a device representing the full calculation.

As an example of the "delta-function method," consider a problem which generalizes the preceding example in two ways: The discrete element is placed at the point $x=d$ rather than at the origin, and the continuous element is gamma distributed with parameters λ and k (as in Section 4.2). Then the delta-function representation of the distribution is

$$\alpha\delta(x-d)+(1-\alpha)\frac{\lambda^k}{\Gamma(k)}e^{-\lambda x}x^{k-1},$$

and the Laplace transform would be

$$\alpha e^{-ds}+(1-\alpha)\left(\frac{\lambda}{\lambda+s}\right)^k,$$

which reduces to the earlier result when $d=0$, $k=1$. The student should, as an exercise, obtain the Laplace–Stieltjes transform by the distribution function formula of Theorem 1.

[†]See, for example, BREMERMANN, HANS (1965), *Distributions, Complex Variables and Fourier Transforms*, Addison-Wesley, Reading, Massachusetts.

It is important to emphasize notation and terminology. Consider, for a function $g(x)$, the two transforms

$$\alpha(s) = \int_0^\infty e^{-sx} g(x) dx,$$

$$\beta(s) = \int_0^\infty e^{-sx} dg(x).$$

The transform $\alpha(s)$ is called the Laplace transform of $g(x)$, and is denoted by $\mathcal{L} g(x)$; the transform $\beta(s)$ is called the Laplace–Stieltjes transform of $g(x)$ and, when $g(x)$ is differentiable, is equal to $\mathcal{L} g'(x)$, which, as will be shown in the next section, is equal to $s\mathcal{L} g(x) - g(0+)$. When $g(x)$ is a continuous distribution function, $g(0+) = 0$, so that in this case the Laplace transform $\alpha(s)$ and the Laplace–Stieltjes transform $\beta(s)$ are connected by the simple formula $\alpha(s) = s\beta(s)$. In the purely discrete case, it has been shown that the Laplace–Stieltjes transform reduces to the probability generating function with argument e^{-s}. For distributions which are mixed discrete and continuous, only the Laplace–Stieltjes transform is defined, although the delta-function format for its calculation is based on an analogy with the Laplace transform.

4.11. Properties of the Laplace Transform

Some properties of the Laplace transform are simple variants of properties of the probability generating function and are proved in a similar, direct fashion: If $\phi(s) = \mathcal{L} f(x)$, with $E(X) = \mu$, then

$$\mu = -\phi'(0) \tag{25}$$

and

$$\text{var}(X) = \phi''(0) - [\phi'(0)]^2. \tag{26}$$

Similarly, if $\psi(s) = \mathcal{L} g(x)$,

$$\phi(s)\psi(s) = \mathcal{L}[f(x) * g(x)]. \tag{27}$$

Other properties refer simply to a change of variables and are proved directly from the definition:

$$\mathcal{L}g(ax)=\frac{1}{a}\psi\left(\frac{s}{a}\right), \qquad a>0, \tag{28}$$

$$\psi(as)=\mathcal{L}\frac{1}{a}g\left(\frac{x}{a}\right), \qquad a>0, \tag{29}$$

$$\mathcal{L}g(x-a)=e^{-as}\psi(s), \qquad g(x)=0,\ x<0,\ a>0, \tag{30}$$

$$\mathcal{L}g(x+a)=e^{as}\psi(s)-\int_0^a e^{-sx}g(x)dx, \qquad a>0, \tag{31}$$

$$\mathcal{L}e^{-ax}g(x)=\psi(s+a). \tag{32}$$

Still other properties are based on differentiation or integration of the variables x or s and are proved in a straightforward way. However, it must be borne in mind that while the existence of a Laplace transform for dg/dx implies that $\mathcal{L}g(x)$ exists, the converse is not true. For example, $\mathcal{L}\log x= -(1/s)\log s$, but $\mathcal{L}(1/x)$ does not converge. Thus the following formulas are applicable when the quantities involved exist:

$$\mathcal{L}\frac{d^n g(x)}{dx^n}=s^n\psi(s)-\sum_{j=0}^{n-1}s^j\frac{d^{n-j-1}g(0+)}{dx^{n-j-1}}, \tag{33}$$

where the symbol $g(0+)$ means that the differentiation is to be performed and the result evaluated as $x\to 0$ from the right. This can be different from the functional value $g(0)$, but need not be so. If, for example, $g(x)=F(x)$, a distribution function, then $F(0)=0$, while $F(0+)$ would be $P(X=0)$ if there was a discrete component at the origin. Similarly,

$$\mathcal{L}\int_0^x\int_0^x\cdots\int_0^x g(u)(du)^n=\frac{\psi(s)}{s^n}, \tag{34}$$

$$\mathcal{L}x^n g(x)=(-1)^n\frac{d^n\psi(s)}{ds^n}, \tag{35}$$

$$\mathcal{L}x^{-n}g(x)=\int_s^\infty\int_s^\infty\cdots\int_s^\infty\psi(u)(du)^n. \tag{36}$$

The special cases $n=1$ of the last four formulas will be of primary interest and, for reference, are given separately:

$$\pounds g'(x)=s\psi(s)-g(0+),\tag{37}$$

$$\pounds \int_0^x g(u)\,du=\frac{\psi(s)}{s},\tag{38}$$

$$\pounds xg(x)=-\psi'(s),\tag{39}$$

$$\frac{\pounds g(x)}{x}=\int_s^\infty \psi(u)\,du.\tag{40}$$

The properties discussed thus far are relatively simple and can be established quite easily from the definitions. There are, however, two important categories of results which are more difficult: (i) limiting theorems, and (ii) inversion theorems. In each case there is a substantial body of purely theoretical analysis, establishing exactly the conditions which are necessary and sufficient. The proofs given here apply only to those situations which are needed in the sequel, although more general theorems are stated without proof.

Limit Theorems[†]

Theorem 1. If $g(x)$ and $g'(x)$ are Laplace transformable, with $\pounds g(x)=\psi(s)$, then

$$\lim_{s\to\infty} s\psi(s)=\lim_{x\to 0+} g(x).$$

Proof. By Eq. (37),

$$\int_0^\infty e^{-sx}g'(x)\,dx=s\psi(s)-g(0+).\tag{41}$$

Since s is independent of the variable of integration, the left side vanishes in the limit and the theorem is proven.   □

[†]Often called *Abelian theorems* or *Tauberian theorems*.

Theorem 2. If $g(x)$ and $g'(x)$ are Laplace transformable and $\lim_{x \to \infty} g(x)$ exists, then

$$\lim_{s \to 0+} s\psi(s) = \lim_{x \to \infty} g(x).$$

Proof. Letting $s=0$ in Eq. (41), the left side becomes

$$\int_0^\infty g'(x)dx = \lim_{x \to \infty} \int_0^x g'(u)du$$

$$= \lim_{x \to \infty} \left[g(x) - g(0) \right].$$

Equating this result to the right side of Eq. (41), it is clear that the quantity $g(0)$ cancels, since it is independent of x and s, and thus the theorem follows. $\qquad \square$

Corollary 1. If $g(x) = f(x)$, a density function, with Laplace transform $\phi(s)$, then

$$\lim_{s \to 0+} s\phi(s) = 0.$$

Corollary 2. If $g(x) = F(x)$, a distribution function with Laplace transform $\Phi(s)$, then

$$\lim_{s \to 0+} s\Phi(s) = 1.$$

Corollary 3. If the Laplace–Stieltjes transform $\sigma(s)$ of a distribution function exists, then

$$\lim_{s \to 0+} \sigma(s) = 1.$$

Proof. This corollary combines the preceding one with Theorem 1 of Section 4.10. $\qquad \square$

Corollary 4. Under the assumptions of the preceding corollary,

$$\lim_{s \to \infty} \sigma(s) = F(0+).$$

In a similar manner, the properties given so far, which apply to Laplace transforms, could be adapted to Laplace–Stieltjes transforms without difficulty. However, it is usually more convenient to reduce the typical Laplace–Stieltjes transform to a Laplace transform by using Theorem 1 of Section 4.10.

The more general theorems, which will not be proven, are as follows: If the various quantities defined exist, then

$$\lim_{x \to 0+} \frac{g(x)}{x^n} = \lim_{s \to \infty} \frac{s^{n+1}\psi(s)}{n!}$$

and

$$\lim_{x \to \infty} \frac{g(x)}{x^n} = \lim_{x \to 0+} \frac{s^{n+1}\psi(s)}{n!}.$$

With all limit theorems it is necessary to take care that the various limits exist, since the existence of one limit does not guarantee that of another. For example, with $g(x) = \sin x$, $\phi(s) = (1+s^2)^{-1}$, and $\lim_{s \to 0+} s\phi(s) = 0$, but $\lim_{x \to \infty} g(x)$ does not exist.

The student should investigate the various limiting formulas for a simple probability distribution, say, $f(x) = \lambda e^{-\lambda x}$, $F(x) = 1 - e^{-\lambda x}$. This distribution could then be compared with the mixed discrete and continuous distribution used as the example in Section 4.10.

4.12. Laplace Inversion

Laplace transforms were originally used, and are still primarily used, to solve differential equations by replacing nth derivatives by nth powers according to Eq. (33) of Section 4.11. In probability, however, the principal use of Laplace transforms is to replace n-fold convolutions by nth powers according to Eq. (27) of Section 4.11. In either situation it may be desirable, once the problem has been solved in terms of the Laplace transform, to obtain the Laplace inverse, e.g., in the case of probability, the distribution function. Unfortunately, the best-known and most general Laplace inversion formula,

$$g(x) = \frac{1}{2\pi i} \lim_{R \to \infty} \int_{c-iR}^{c+iR} e^{sx}\psi(s)\,ds,$$

involves integration in the complex plane. Moreover, this integration cannot be performed, except numerically, in many quite ordinary cases. There are several ways out of this difficulty which confronts nearly every user of Laplace transforms. One possibility is to use tables of Laplace transforms or inverse Laplace transforms. For example, in the volume *Tables of Integral Transforms, Vol. 1,*[†] formulas are given for 785 Laplace transforms and 687 inverse Laplace transforms, as well as a number of general formulas of the type treated in Section 4.11.

In this textbook, a simpler formula, which can be proved without too much difficulty, will be adequate for the problems encountered. This formula is based on a property of the gamma distribution.

Consider a gamma distribution with parameters λ and k, so that the mean μ and variance v satisfy

$$\lambda\mu = k, \qquad v\lambda^2 = k.$$

Define a limit so that the mean remains fixed and the variance approaches zero; this can be done by replacing the parameters λ and k by μ and k, so that the density function becomes

$$\frac{(k/\mu)^k}{\Gamma(k)} e^{-kx/\mu} x^{k-1}$$

and the limit is obtained as $\lim_{k\to\infty}$. It is intuitively clear (and can be proven rigorously) that in the limit, the gamma distribution becomes a causal distribution with probability unity at the value μ and, moreover, that any bounded continuous function $g(x)$, integrated with respect to the gamma distribution, becomes $g(\mu)$, that is,

$$\lim_{k\to\infty} \int_0^\infty \frac{(k/\mu)^k}{\Gamma(k)} e^{-kx/\mu} x^{k-1} g(x)\,dx = g(\mu). \tag{42}$$

Theorem 1. If $f(x)$ is a wholly continuous density function with Laplace transform $\phi(s)$, then

$$f(x) = \lim_{k\to\infty} \frac{(-1)^{k-1}}{\Gamma(k)} \left(\frac{k}{x}\right)^k \frac{d^{k-1}}{ds^{k-1}} \phi\left(\frac{k}{x}\right). \tag{43}$$

[†]ERDELYI, A., ed. (1954), McGraw-Hill, New York.

Proof. Differentiating the Laplace transform $k-1$ times gives

$$\frac{d^{k-1}\phi}{ds^{k-1}} = \int_0^\infty (-u)^{k-1} e^{-su} f(u)\, du$$

and, with the variable s replaced by k/x,

$$\frac{d^{k-1}\phi(k/x)}{ds^{k-1}} = \int_0^\infty (-1)^{k-1} u^{k-1} e^{-ku/x} f(u)\, du.$$

Substituting this result into the right side of Eq. (43) gives

$$\lim_{k\to\infty} \int_0^\infty \frac{(k/x)^k}{\Gamma(k)} e^{-ku/x} u^{k-1} f(u)\, du,$$

which, by Eq. (42), is equal to $f(x)$, as stated by the theorem. □

This Laplace inversion theorem depends for its success on the ability to differentiate the Laplace transform $k-1$ times. If, for example, $\phi(s) = \lambda/(\lambda+s)$, then $\phi^{k-1}(s) = (-1)^{k-1}(k-1)!\lambda(\lambda+s)^{-k}$ and the student can show that the theorem yields $f(x) = \lambda e^{-\lambda x}$.

On the other hand, in the mixed discrete and continuous example given in Section 4.10, the theorem could not be used until the two components were separated in the Laplace transform. When the constant portion is isolated, this corresponds to the discrete component, and then the remaining portion, corresponding to the continuous component, can be successfully inverted. However, the discrete component need not correspond to a constant in the Laplace transform; the student should investigate a distribution with discrete components at the origin and at the value $x = 1$, together with a continuous density.

4.13. Random Sums

The procedure of random summing discussed in Section 3.12 can equally well be applied to continuous and mixed discrete and continuous distributions. Let $S_N = X_1 + X_2 \cdots + X_N$, where the X_j are independently and identically distributed with Laplace transform $\eta(s)$, and let the probability generating function of N (necessarily a discrete variable) be $\beta(s)$.

Then the Laplace transform $\zeta(s)$ of S_N can be written

$$\zeta(s) = \beta[\eta(s)],\tag{44}$$

with the same argument as used in Section 3.12; when the X_j are discrete, the result given in that section follows from

$$\phi(e^{-s}) = \beta[\alpha(e^{-s})].$$

The student will verify that when the X_j are exponential with parameter λ, and N is geometric with parameter ρ, then S_N is exponential with parameter $\lambda(1-\rho)$.

The distribution of the X_j is called the *stopped* distribution and that of N, the *stopping* distribution.

Just as in Section 3.12, the method of marks can be used to obtain the nested probability generating function equation.

An important special case is that obtained when N is Poisson distributed. Then

$$\zeta(s) = \exp\{-\lambda[1 - \eta(s)]\}.\tag{45}$$

The resulting distribution is called a *compound Poisson distribution*.

Recommendations for Further Study

To treat continuous random variables in a completely rigorous manner, it is necessary to face squarely the fact that probability is a measure and introduce some of the machinery of measure theory. A good introduction to this subject is contained in the highly recommended book by Neuts (1973); a more comprehensive and exact treatment is contained in the book of Breiman (1968). For a further study of the Laplace transform, the books of Giffin (1975) and Widder (1941) are recommended.

BREIMAN, LEO (1968), *Probability*, Addison-Wesley, Reading, Massachusetts.

GIFFIN, WALTER C. (1973), *Transform Techniques for Probability Modelling*, Academic Press, New York.

NEUTS, MARCEL F. (1973), *Probability*, Allyn and Bacon, Boston.

WIDDER, D. V. (1941), *The Laplace Transform*, Princeton University, Princeton.

4.14. Problems[†]

1. On a line segment XY, two points are chosen independently and at random. If the points are P and Q, what is the probability that the segments XP, PQ, and QY can form a triangle?

2. A point is chosen at random on the base of an equilateral triangle of side A. Find the density function for the random variable X, defined as the distance from the chosen point to the opposite vertex.

$$Ans. \quad (2x/A)(4x^2 - 3A^2)^{-1/2}, \tfrac{1}{2}\sqrt{3}\, A < x < A$$

3. Consider a random variable with range $(0,2)$ and with the density function $2Kx(2-x)$. Find K and determine whether the distribution is symmetrical about the mean.

4. A point is chosen at random in the unit square. What is the probability that the point lies within the triangle formed by the y axis, the diagonal $x=y$, and the horizontal $y=1$? What would the probability be if it were given that the point fell within the triangle formed by the coordinate axes and the diagonal $x+y=1$?

5. A random variable with range $(0,1)$ has density function $kx^2(1-x^3)$. Find the value of k and the expectation of the variable. $\qquad\qquad Ans.\ E(x)=\tfrac{9}{14}$

6. A random variable with range $(2,5)$ has density function $k(1+x)$. Find (i) k, (ii) $P(X>3)$, (iii) $E(X)$.

7. Let X denote the lifetime for a type of light bulb, where

$$P(X<x)=\int_0^x Ct^2 e^{-pt}dt.$$

Express C in terms of p and find $E(X)$.

8. In Problem 7, find the probability that a bulb which has been in service for two time units and is still burning will continue to burn for at least three more time units.

9. In Problem 7, what would be the mean life expected of a bulb which has burned for two time units? How much longer would such a bulb be expected to burn?

10. A point is chosen at random inside the unit circle. Find the density function for X, where X is (i) the distance from the point to the center of the circle, (ii) the square of the distance from the point to the center of the circle.

11. A given amount of time T is to be spent searching for oil in two places. If it is in the first place, and if a time t is spent searching there, the probability of finding it is $1-e^{-kt}$, where k is a positive constant. The remaining time $T-t$ is spent

---

[†]Problems 13, 28, and 40 are taken from Lindley (1965) (see p. 132) with kind permission of the publisher.

searching in the second place, and if the oil is there, the probability of finding it is $1-e^{-k(T-t)}$. Given that the probability of oil being in the first place is p, and in the second place, $1-p$, what is the probability that oil will be found? How should the time be divided between the two places to maximize the probability of discovery?

12. A point P is chosen at random on the diameter of a semicircle of unit radius and a perpendicular is drawn to meet the semicircle in Q. Find the expected length of PQ. Another point P' is chosen at random on the circumference of the semicircle and a perpendicular to the diameter is drawn through P' to meet the diameter in Q'. Find the expected length of $P'Q'$ and explain why the two results do not agree.

13. A point X is chosen at random on the unit line segment PQ. What is the expected area of the rectangle with sides PX and XQ? What is the probability that the area is greater than one-half?

14. Suppose a random variable has a density function which is symmetric about the value $X=c$. Show that the density function of $aX+b$ is symmetric about the value $ac+b$.

15. Suppose a random variable X has range $(0, \infty)$ and a random variable Y has range $(-X, +X)$, with joint density

$$f(x, y)=k(x^2-y^2)e^{-x}.$$

Find the value of k, the marginal and conditional densities, and the expectations of X and Y.

16. Suppose the range for the distribution of X and Y is the unit circle, with joint density

$$k(1-x^2-y^2).$$

Find the value of k, the marginal and conditional densities, and the expectations of X and Y. $Ans.\ \left(\dfrac{8}{3\pi}\right)(1-x^2)^{3/2},\ -1<x<1$

17. Consider independent random variables X and Y with densities $6x(1-x)$ and 1, respectively, each being defined over the range $(0, 1)$. Find the densities of (i) $X+Y$, (ii) $X-Y$.

18. Let X and Y be independently negative exponential with the same parameter λ. Show that the random variables $X+Y$ and X/Y are independent.

19. Suppose random variables X and Y each have range $(0, \infty)$, with joint density function

$$e^{-x/y}y^{-1}e^{-y}.$$

Show that $E(X|Y=y)=y$.

20. If X and Y are independently negative exponentially distributed (with different parameters), find the distribution of min(X, Y). Hint: See Problem 18.

21. If X and Y are independent negative exponentially distributed random variables (with the same parameter), what is the probability that the equation

$$u^2 - 2Xu + Y = 0$$

has real, distinct roots? Note: you need not evaluate the integral in the answer.

22. Let X and Y be independent negative exponentially distributed random variables with the same parameter. Show that $X/(X+Y)$ is rectangularly distributed over the unit interval.

23. Show that the ratio of two independent negative exponential random variables with the same parameter has density function $(1+x)^{-2}$. What is the range?

24. Random variables X and Y have joint density function

$$4y(x-y)e^{-(x+y)}, \qquad 0<x<\infty, 0<y<x$$

Find the marginal and conditional densities.
    Ans. $4ye^{-2y}, 0<y<\infty$ for Y and $(x-y)e^{-x+y}, y<x<\infty$ for $X|Y$.

25. Assuming that the parameters p and q in the beta distribution both lie in the interval $(0, 1)$, show that the density function has a unique maximum and no inflection points.

26. Use the method of marks to obtain the compound Poisson distribution in the case where the stopped distribution is discrete, so that Eq. (45) of Section 4.13 connects probability generating functions rather than Laplace transforms.[†]

27. Apply Eq. (18) of Section 4.8 to the Poisson distribution.

28. Suppose the duration t, in minutes, of long-distance telephone calls made from a certain city is found to be a random variable with distribution function

$$F(t)=0, \qquad \text{for } t<0,$$

$$= 1 - \tfrac{2}{3}e^{-t/3} - \tfrac{1}{3}e^{-[t/3]}, \qquad \text{for } t \geq 0,$$

where $[t/3]$ is the integral part of $t/3$. (i) Sketch the distribution function. (ii) Is the random variable discrete, continuous, or mixed? (iii) What is the probability that the duration of a call will be (a) more than six, (b) less than four, (c) equal to three, (d) between four and seven minutes? (iv) What is the conditional probability that the duration of a call will be less than nine minutes, given that it has lasted more than five minutes?

[†]See RÅDE, LENNERT (1972), On the use of generating functions and Laplace transforms in applied probability theory. *Int. J. Math. Ed. Sci. Technol.* **3**, 25–33.

29. Show that if the Poisson probabilities are *truncated* by normalizing to unity those beyond the value $N-1$, the resulting probabilities can be written

$$\frac{\Gamma(N)e^{-\lambda}\lambda^x}{x!\gamma(N,\lambda)}, \qquad x=N,N+1,N+2,\dots .$$

30. In Eq. (14), show that the moment generating function of the parameter-mixed distribution can be written in terms of the Laplace transformation of $f(x)$.

31. Show that the Laplace transformation of

$$\int_0^x u^{-1}f(u)\,du$$

is given by

$$s^{-1}\int_s^\infty \phi(u)\,du.$$

32. Let X be beta distributed with parameters p and q. Find the density of $1/X-1$.

   Ans. $x^{p-1}(1+x)^{-p-q}/B(p,q), 0<x<\infty$.

33. Show that if X and Y are independent, gamma distributed random variables with parameters λ, k and λ, K, respectively, then $X/(X+Y)$ is beta distributed with parameters k and K.

34. Prove

   (i)
   $$\frac{d^k}{dx^k}\left[e^x\gamma(n,x)\right]=(-1)^k(1-n)_k e^x\gamma(n-k,x),$$

   (ii)
   $$\frac{d^k}{dx^k}\left[e^x\Gamma(n,x)\right]=(-1)^k(1-n)_k e^x\Gamma(n-k,x).$$

35. Prove

   (i)
   $$\frac{d^k}{dx^k}\left[x^{-n}\gamma(n,x)\right]=(-1)^k x^{-n-k}\gamma(k+n,x),$$

   (ii)
   $$\frac{d^k}{dx^k}\left[x^{-n}\Gamma(n,x)\right]=(-1)^k x^{-n-k}\Gamma(k+n,x).$$

36. Prove

$$\frac{d\gamma(n,x)}{dx}=-\frac{d\Gamma(n,x)}{dx}=x^{n-1}e^{-x}$$

37. Prove

(i)
$$\sum_{j=1}^{\infty} \frac{\gamma(j,\lambda)}{\Gamma(j)} = \lambda,$$

(ii)
$$\sum_{j=1}^{\infty} j \frac{\gamma(j,\lambda)}{\Gamma(j)} = \tfrac{1}{2}\lambda^2 + \lambda.$$

38. If X is a continuous random variable with distribution function $F(x)$, show that $E(X) = \int_0^{\infty}[1 - F(x)]\,dx$.

39. Using the substitutions $t = \sin u$, show that

$$B(a,b) = 2\int_0^{\pi/2}(\sin u)^{2a-1}(\cos u)^{2b-1}\,du.$$

40. The probability density of the velocity V of a molecule with mass m in a gas at absolute temperature T is

$$Av^2 e^{-\beta v^2},$$

with range $(0,\infty)$, where $\beta = m/2kT$, k is Boltzmann's constant, and A is chosen for normalization to unity. Find the mean and variance of V.

41. The *chi-square distribution*, useful in statistics, has density function

$$\frac{x^{(n/2)-1}e^{-x/2}}{2^{n/2}\Gamma(\tfrac{1}{2}n)}.$$

Show that this is actually a gamma distribution, and find the values of the parameters.

42. Show that

$$B_x(p,q) + B_{1-x}(q,p) = B(p,q).$$

43. Show that

(i)
$$\int_0^{\pi/2}(\cos u)^r\,du = \frac{\Gamma[\tfrac{1}{2}(r+1)]\,\pi^{1/2}}{2\Gamma(\tfrac{1}{2}r+1)}, \qquad r > -1,$$

(ii)
$$\int_0^{\pi/2}(\sin u)^r\,du = \frac{\Gamma[\tfrac{1}{2}(r+1)]\,\pi^{1/2}}{2\Gamma(\tfrac{1}{2}r+1)}, \qquad r > -1.$$

44. Using the transformation $t=u/(1+u)$, show that

$$B(a,b)=\int_0^\infty \frac{u^{b-1}du}{(1+u)^{a+b}}.$$

45. Show that

$$\int_0^\infty \frac{\Gamma(k,\lambda x)}{\Gamma(k)}dx=\frac{k}{\lambda}.$$

46. Verify Eq. (8).

5

Continuous Time Processes

5.1. Introduction and Notation

There is a fundamental difference between the mathematical formulation of a discrete time stochastic process and a continuous time stochastic process. In discrete time, it is necessary to specify only the mechanism for transition from one state to another, and of course the initial state (distribution) of the system. For Markov chains, this consists of the transition matrix and the initial vector. Everything about the chain can, in principle, be deduced from this matrix and vector.

For a continuous time process, on the other hand, it is insufficient to specify the mechanism for changes of state. Changes of state can occur at any instant of a time continuum, and it is therefore also necessary to specify *when* the changes occur. A process which oscillates between two states, for example, might jump from one state to the other every hour on the hour, or might remain in one state for a time which is a random variable. The two processes would clearly be quite different. In discrete time, a two-state process would be defined simply by giving the probabilities of a state change. The question of when the discrete time process can change states is not relevant.

Every continuous time process, therefore, has associated with it a process consisting of those instants of time when transitions take place. This is called its associated *point process*, since the instants are a sequence of points on the time axis. For a given point process there can be defined many different transition schemes. The simplest is the one which begins at zero and increases by one unit at each transition point, and thus counts the number of transition points. Such a process is called the *counting process* of the point process. A counting process consists of a random variable $X(t)$, which is the number of points of the point process in the interval $(0, t]$.

The transition times will be denoted by $\tau_0 = 0$, $\tau_1, \tau_2, \ldots$, and the intervals between consecutive transition times by

$$\sigma_n = \tau_{n+1} - \tau_n, \qquad n = 0, 1, 2, \ldots.$$

These will be called the *gaps* of the point process. When the point process is considered to be the transition points of a continuous time, stochastic process, the gaps are sometimes also called the *sojourn times* of the process.

It is clear that a point process can be specified either by giving a mechanism for determining the points τ_n or, alternatively, by giving a mechanism for determining the gaps σ_n. Sometimes it will be convenient to do one, and sometimes the other. Indeed the relationship between the two formulations will play an important role in the theory.

In simple cases, the process which specifies the gaps or points of transition of a stochastic process is independent of the process which specifies the transitions. This is by no means always the case. There are situations in which the nature of each succeeding gap will depend on the state of the system at the beginning (or even at the end) of that gap.

Notation

Let $X(t)$ denote the state of the system at time t. Then the double subscript probability with two time values

$$p_{xy}(u, t) = P\big[X(u+t) = y \,|\, X(u) = x\big]$$

represents the probability of a transition from state x to state y in the interval $(u, u+t)$.

A *Markov process* will be defined by negative exponential sojourn times, so that $p_{xy}(u, t)$ will be independent of u, by the zero-aging property of the negative exponential distribution. In this case the extra time variable is omitted, writing

$$p_{xy}(t) = P\big[X(u+t) = y \,|\, X(u) = x\big]. \tag{1}$$

For a Markov process, the exponential parameter is not necessarily the same for each gap. The unconditional distribution will also be denoted by the letter p, but with a single subscript (Markov or not):

$$p_x(t) = P[X(t) = x]. \tag{2}$$

Unless otherwise specified, it will be assumed that the state space of the process is the non-negative integers $0, 1, 2, \ldots$.

As in Chapter 3, the initial distribution will be denoted by the letter a:

$$a_x = P[X(0) = x].$$

It is still true, as by analogy to Chapter 3, that

$$p_y(t) = \sum_x a_x p_{xy}(0, t),$$

or, in the Markov case,

$$p_y(t) = \sum_x a_x p_{xy}(t). \tag{3}$$

A counting process will be characterized by

$$X(t + u) = X(u) + N(t),$$

where $N(t)$ is the number of transitions in time t. Thus, for a counting process,

$$p_{xy}(u, t) = P[N(t) = y - x \mid X(u) = x]. \tag{4}$$

Also by analogy to Chapter 3, when an equilibrium (or steady-state or stationary) distribution is obtained by letting $t \to \infty$, it will be denoted by p_x:

$$\lim_{t \to \infty} p_x(t) = p_x.$$

Additional notation, especially for non-Markov cases, will be developed as needed.

The Chapman–Kolmogorov equations for the continuous time case are analogous to those for the discrete time case, and they merely formalize the fact that to pass from one state to another, it is possible to pass through a third state:

$$p_{xy}(t_1 + t_2) = \sum_z p_{xz}(t_1) p_{zy}(t_2). \tag{5}$$

5.2. Renewal Processes

A point process which has identically equidistributed gaps, with the same parameter(s), is called a *renewal process*. In mathematical terms, a renewal process is specified by the condition

$$P(\sigma_n < x) = G(x), \qquad n = 0, 1, 2, \ldots. \tag{6}$$

In general, it will be assumed that the gap distribution is entirely continuous, so that the *gap density* $g(x) = G'(x)$ exists. However, the limiting case of equally spaced transition points, which corresponds to the causal gap distribution, will also be discussed.

The symbols γ and v will be used to denote the gap mean and variance, so that $\lambda = 1/\gamma$ is the transition-point frequency parameter.

Three special renewal processes are especially important:

The Erlang Process, with gamma-distributed gaps:

$$g(x) = \frac{\mu^k}{\Gamma(k)} e^{-\mu x} x^{k-1}.$$

The Poisson Process, an Erlang process with $k = 1$, that is, negative exponentially distributed gaps, with fixed parameter μ. (This is a special case of a Markov process, in which each exponential gap has the same parameter.)

The Deterministic Process, with equally spaced transition points. This can be obtained as a limiting Erlang process with the mean fixed and the variance approaching zero, i.e., $\mu \to \infty$ and $k \to \infty$, so that μ/k remains fixed.

Renewal processes are important in their own right, not only as point processes furnishing transition points for more complex processes. There is, in fact, a general theory of renewal processes, complete with characteristic terminology and results. It is worthwhile to discuss this theory and to mention some of the alternative terminology.

The original idea, which lead to the word "renewal," was that of replacing ("renewing") items subject to eventual failure. The transition points correspond to the instants when the failed item is replaced, and the gaps correspond to the *lifetimes* of the items. Thus the gap density $g(x)$ is called the *lifetime* density, and its tail $1 - G(x)$ is called the *survivor function*, since it gives the probability that an item aged x is still functioning.

A function of interest in renewal theory is the *failure rate* $f(x)$, which represents the probability of the immediate failure of an item aged x, or, in other words, the probability of a transition point, given time x since the last transition point. The relationship between $f(x)$ and $g(x)$ is easy to obtain; if X is the gap length, then

$$f(x) = \lim_{\Delta x \to 0} \frac{P(x < X < x + \Delta x \mid x < X)}{\Delta x}$$

$$= \lim_{\Delta x \to 0} \frac{P(x < X < x + \Delta x)}{\Delta x} \frac{1}{P(x < X)}$$

$$= \frac{g(x)}{1 - G(x)}, \tag{7}$$

or, alternatively,

$$f(x) = -\frac{d}{dx} \log[1 - G(x)]. \tag{8}$$

The student should show that $g(x)$ can be expressed in terms of $f(x)$,

$$g(x) = f(x) \exp\left(-\int_0^x f(u)\,du\right), \tag{9}$$

and hence that a renewal process has constant failure rate if and only if it is a Poisson process. This property corresponds to the lack of aging property of the negative exponential distribution. According to either interpretation, a transition point is equally likely to occur at any instant, quite independently of the length of time since the last transition point.

In a renewal process, if $f(x)$ is an increasing function, the process is said to have *positive aging*, if decreasing, *negative aging*. The student should examine the relationship between $f(x)$ and the parameters of the gamma distribution which define an Erlang process (Problem 28). What does this imply for the aging of a deterministic process?

5.3. The Poisson Process

A Poisson process is both Markov and renewal, that is, it is defined by negative exponential gaps (Markov) with the same parameter (renewal). The counting distribution for a Poisson process is a Poisson distribution. A proof of this fact depends on the Poisson–gamma relationship of Section 4.5. Let $N(t)$ be the number of points in the interval $(0, t]$, so that $N(t)$ represents the state of the Poisson process at time t. Then

$$P(N(t)<x)=P(\tau_x>t)$$

$$=P(X_0+X_1+\cdots+X_{x-1}>t), \qquad (10)$$

exactly as in Section 4.5.

Another approach to the counting distribution for a Poisson process is by means of the infinitesimal properties of the negative exponential distribution. Writing out the power series for $e^{-\lambda}$, it is easy to see that the probability of one renewal point in an interval Δt is $\lambda\Delta t+o(\Delta t)$, the probability of no renewal point is $1-\lambda\Delta t+o(\Delta t)$, and that of more than one point is $o(\Delta t)$.[†] Based on this information, let a function $L(t)$ satisfy the following system of axioms:

Axiom 1. $P(L(\Delta t)=0)=1-\lambda\Delta t+o(\Delta t)$.

Axiom 2. $P(L(\Delta t)=1)=\lambda\Delta t+o(\Delta t)$.

Axiom 3. $P(L(\Delta t)>1)=o(\Delta t)$.

Axiom 4. The events defined by nonoverlapping intervals are independent.

Let $p_x(t)=P(L(t)=x)$. Then

$$p_x(t+\Delta t)=P(L(t+\Delta t)=x)$$

$$=\sum_{j=0}^{x} P(L(t)=j)P(L(\Delta t)=x-j)$$

$$=(1-\lambda\Delta t)p_x(t)+\lambda\Delta t p_{x-1}(t)+o(\Delta t).$$

Forming the derivative with respect to t in the usual way gives

$$p_x'(t)=-\lambda p_x(t)+\lambda p_{x-1}(t), \qquad x=1,2,3,\ldots. \qquad (11)$$

[†] In usual mathematical notation, $o(h)$ denotes a function having the property $\lim_{h\to 0}[o(h)/h] = 0$.

When $x=0$, the argument needs to be modified. The student should fill in the details to show that

$$p_0'(t) = -\lambda p_0(t). \tag{12}$$

Using the generating function technique, Eqs. (11) and (12) can be combined in the following form:

$$\phi'(s, t) = -\lambda(1-s)\phi(s, t), \tag{13}$$

where

$$\phi(s, t) = \sum_{j=0}^{\infty} p_j(t)s^j.$$

The solution to the differential equation (13) is

$$\phi(s, t) = \exp\left[-(1-s)\lambda t\right],$$

which is the probability generating function for a Poisson distribution with parameter λt.

This result can be interpreted in the following way: If Axioms 1–4 describe the placement of transition points with a counting function $L(t)$, then the resulting counting process is Poisson.

The theorem also shows that the probability of x points in time t is Poisson distributed with parameter λt—that the interval need not begin at the origin. This should also be clear from the lack of aging of the negative exponential distribution, so that each point is equivalent to every other point as the first point of a counting interval. It is easy to show (again referring to the Poisson–gamma relationship) that negative exponential gaps are implied by Poisson counts. Therefore, for any point process other than the Poisson process, the probability of x points in an interval of length t will depend on where the beginning of the interval is situated with respect to the last point of the process. Thus only in connection with the Poisson process is it correct to speak of the counting distribution without specifying the beginning of the counting interval.

The argument leading to the Poisson distribution in this section makes use of differential difference equations: Eq. (11) is a differential equation in the continuous variable t and also a difference equation in the discrete variable x. By using probability generating functions, the effect of the differences in x was removed and the resulting differential equation was solved. This technique will be frequently used.

5.4. Two-State Processes

A two-state continuous time process models phenomena which switch from time to time from one state to another: "on" to "off," "red" to "green," "up" to "down," and back again. In contrast with the discrete time process, no possibility exists for transition from one state to itself, and so the formulation consists essentially of a description of the point process of transition instants. For a traffic light, these might be lattice points.

For a Markov process, the gaps would be negative exponential. The Markov property for a two-state process could therefore be expressed in terms of infinitesimals, just as the Poisson process was in Section 5.3: If

$$a(t)=P(\text{transition from state 0 to state 1 in time } t),$$

$$b(t)=P(\text{transition from state 1 to state 0 in time } t),$$

then

$$a(\Delta t)=\lambda t+o(\Delta t),$$

$$b(\Delta t)=\mu t+o(\Delta t).$$

Let

$$p_1(t)=P(\text{process is in state 1 at time } t)$$

$$=1-p_0(t).$$

Using the same kind of argument as in Section 5.3, the differential equation in $p_1(t)$ can be built up by conditioning on the state of the system at time t as follows:

$$p_1(t+\Delta t)=p_1(t)(1-\mu\Delta t)+p_0(t)\lambda\Delta t+o(\Delta t).$$

The differential equation is then

$$p_1'(t)=-(\lambda+\mu)p_1(t)+\lambda, \tag{14}$$

and this can be solved by using a standard formula[†] to yield

$$p_1(t) = \frac{\lambda}{\lambda+\mu}(1+Ce^{-(\lambda+\mu)t}).$$

Here C is a constant which can be evaluated by making some assumptions regarding the initial state of the system. The student can work out the details under the assumptions that (i) the system begins in state 0, (ii) the system begins in state 1, (iii) the system has probability π of beginning in state 0 and probability $1-\pi$ of beginning in state 1.

It will be noted that, independent of the initial state of the system, the probabilities evaluated as $t\to\infty$ are proportional to the coefficients λ and μ. The meaning of this result will be discussed in more detail in Section 5.8.

5.5. Markov Processes

A continuous time Markov process is defined by the negative exponentially distributed sojourn times, in which the parameter λ_{xy} can depend both on the state x from which the transition takes place and the state y to which the transition takes place. Since this formulation depends on the assumption that there is a transition (so that the gap between transitions actually ends), the formal definition applies only when $x\neq y$. If $p_{xy}(t)$ denotes the probability that, beginning at time zero in state x, the system is in state y at time t,

$$p_{xy}(t) = P(X(t)=y \mid X(0)=x),$$

then, by definition of a Markov process,

$$p_{xy}(\Delta t) = \lambda_{xy}\Delta t + o(\Delta t), \qquad x\neq y. \tag{15}$$

An expression for $p_{xx}(\Delta t)$ is not difficult to obtain. Since

$$\sum_k p_{xk}(\Delta t) = 1,$$

it follows that

$$p_{xx}(\Delta t) = 1 - \sum_{k\neq x} p_{xk}(\Delta t)$$

$$= 1 - \Delta t \sum_{k\neq x} \lambda_{xk} + o(\Delta t). \tag{16}$$

[†] If $y' + Py = Q$, then $y = e^{-\int P\,dx}(\int e^{\int P\,dx}Q\,dx + C)$.

With this system of notation for the *infinitesimal probabilities* $p_{xy}(\Delta t)$, which incorporates the Markov property, it is possible to compute, in the general case, the differential difference equation in much the same style as has already been done for the Poisson and the two-state processes. The basic idea is to divide the time period $(t, t+\Delta t)$ into periods of length t and Δt and form the derivative:

$$p_{xy}(t+\Delta t) = \sum_k p_{xk}(t)p_{ky}(\Delta t)$$

$$= \sum_{k\neq y} p_{xk}\lambda_{ky}\Delta t + p_{xy}(t)\left(1 - \Delta t \sum_{k\neq y} \lambda_{yk}\right) + o(\Delta t), \quad (17)$$

$$p'_{xy}(t) = \sum_{k\neq y} p_{xk}(t)\lambda_{ky} - p_{xy}(t) \sum_{k\neq y} \lambda_{yk}. \quad (18)$$

It is now possible to express the coefficients λ_{xy} in terms of the derivatives of $p_{xy}(t)$ at the origin. Setting $t=0$ when $x\neq y$ gives

$$p'_{xy}(0) = \lambda_{xy}, \quad x\neq y.$$

Setting $x=y$ and then $t=0$ gives

$$p'_{xx}(0) = - \sum_{k\neq x} \lambda_{xk}.$$

It will be noted that λ_{xx} is not yet defined; now it will be taken to be $p'_{xx}(0)$. The system of differential equations (18) is known as the *forward differential equation* system of the process. An analogous system of *backward differential equations* can be obtained by beginning with t and Δt interchanged.

In either case, the solutions of the differential difference equations are obtainable only in certain special cases. Two of these have already been investigated:

The Poisson Process

$$\lambda_{xx} = -\lambda, \quad \lambda_{x,x+1} = \lambda, \quad \lambda_{xy} = 0, \quad y\neq x, x+1.$$

The Two-State Process

$$\lambda_{0,1} = \lambda, \quad \lambda_{1,0} = \mu.$$

The student should work out the details of these processes as special cases of the Markov process.

Another interesting special case is the counting distribution for a Markov point process, in which $p_x(t)$ is the probability of x transitions in time t. This is known as the *pure birth process* when the sojourn parameter depends only on the initial state of the system, $\lambda_{xy} = \lambda_x$.

In this process,

$$p_0(\Delta t) = 1 - \lambda_x \Delta t + o(\Delta t),$$

$$p_1(\Delta t) = \lambda_x \Delta t + o(\Delta t),$$

$$p_x(\Delta t) = o(\Delta t), \qquad x = 2, 3, 4, \ldots.$$

The calculations leading to the differential difference equations are much the same as in the Poisson case, except for the fact that the λ_x are not all equal and, in fact, mesh with the state x of the system.

Beginning with

$$p_x(t + \Delta t) = p_x(t)[1 - \lambda_x \Delta t] + p_{x-1}(t)\lambda_{x-1}\Delta t + o(\Delta t),$$

straightforward calculations (which the student should perform) yield

$$p_x'(t) = -\lambda_x p_x(t) + \lambda_{x-1} p_{x-1}(t), \qquad x = 1, 2, 3, \ldots,$$

$$p_0'(t) = -\lambda_0 p_0(t).$$

(19)

It is not easy to solve these equations, in general. However, when the birth rate λ_x is linear ($\lambda_x = \lambda x$), a closed form solution can easily be obtained. The linear birth process is called a *Yule process*, and it has often been applied to problems where the probability of an addition to the population is proportional to the size of the population. For the Yule process, it can be shown by induction, or by use of probability generating functions, that the distribution of population size at time t, assuming an initial population of N members, is negative binomial over the values $N, N+1, N+2, \ldots$ (compare Section 1.10), with parameters

$$n = N, \qquad p = 1 - e^{-\lambda t}.$$

The student should carry out the calculations leading to this conclusion, and, by investigation of the negative binomial mean for this case, show that the population mean is exponential with time.

5.6. Equilibrium

In several models the time derivatives of the state probabilities have been calculated. When the probabilities do not change with time, the system is said to be in equilibrium. In some cases, finite equilibrium probabilities exist and can be obtained by setting the time derivatives equal to zero. In other systems, no such equilibrium exists, and the probabilities of every state approach zero as time goes to infinity. The pure birth model, for example, providing as it does for additions to the population but not for subtractions, would be expected not to have an equilibrium, and indeed it does not. Setting $p'_x(t)=0$ in Section 5.5 yields only $p_x(t)=0$ for all x.

As an example of a Markov process with an equilibrium, consider the *birth-and-death process*. This is a Markov process which can at each transition point go only into the next higher or next lower state. From state x, the probability of transition to state y is, similarly to the pure birth process, given by the infinitesimal probabilities

$$p_{x,x+1}(\Delta t)=\lambda_x\Delta t+o(\Delta t).$$

Another sequence of parameters μ_x governs the "death" aspect of the process, i.e., transitions form x to $x-1$:

$$p_{x,x-1}(\Delta t)=\mu_x\Delta t+o(\Delta t).$$

Events like the occurrence of a birth and a death in time Δt have probability $o(\Delta t)$, and there is no change with probability

$$p_{xx}(\Delta t)=(1-\lambda_x\Delta t-\mu_x\Delta t)+o(\Delta t).$$

These transition probabilities can be applied to the calculation of $p'_x(t)$, as follows:

$$p_x(t+\Delta t)=p_x(t)(1-\lambda_x\Delta t-\mu_x\Delta t)$$

$$+\lambda_{x-1}p_{x-1}(t)\Delta t+\mu_{x+1}p_{x+1}(t)\Delta t+o(\Delta t),$$

leading to

$$p'_x(t)=-(\lambda_x+\mu_x)p_x(t)+\lambda_{x-1}p_{x-1}(t)+\mu_{x+1}p_{x+1}(t), \qquad x=1,2,3,\ldots,$$

$$(20)$$

with the special case

$$p_0'(t) = -\lambda_0 p_0(t) + \mu_1 p_1(t).$$

Assuming that an equilibrium exists, the equilibrium probabilities can be found by setting the left sides of these equations equal to zero. Then

$$p_1 = \frac{\lambda_0}{\mu_1} p_0,$$

$$p_2 = \frac{\lambda_0 \lambda_1}{\mu_1 \mu_2} p_0,$$

$$\vdots \qquad \vdots \qquad\qquad\qquad (21)$$

$$p_x = \frac{\lambda_0 \lambda_1 \cdots \lambda_{x-1}}{\mu_1 \mu_2 \cdots \mu_x} p_0,$$

$$\vdots \qquad \vdots$$

The necessary condition for the equilibrium is therefore that this sequence sums to $1 - p_0$, i.e., that the series

$$p_0^{-1} = \sum_{j=0}^{\infty} \frac{\lambda_0 \cdots \lambda_j}{\mu_1 \cdots \mu_{j+1}} + 1 \qquad (22)$$

converges.

A special case of the birth-and-death process is the Markov queue: $\lambda_x = \lambda$, $\mu_x = \mu$. The random variable represents the number in the system (queueing and being served), while λ represents the rate of arrival to the system, and μ, the rate of service. Then, writing $\rho = \lambda/\mu$,

$$p_0 = 1 - \rho, \qquad p_x = (1 - \rho)\rho^x,$$

a geometric distribution.

For a queue of this type, the variance is always larger than the mean and approaches infinity with saturation ($\rho \to 1$) more rapidly. It is therefore quite sensible to "come back later" when the queue seems too long to be tolerable.

Another special case of interest is the linear birth-and-death process, specified by the parameter values $\lambda_x = \lambda x$, $\mu_x = \mu x$. Then Eqs. (20) can be written

$$p_0'(t) = \mu p_1(t),$$

$$p_x'(t) = -(\lambda + \mu)xp_x(t) + \lambda(x-1)p_{x-1}(t) + \mu(x+1)p_{x+1}(t),$$

(23)

with the equilibrium solution $p_0(\infty) = 1$. Thus the probability of extinction is 1, and the zero state is absorbing. Some further results on the linear birth-and-death process are given in problems 36–38 of Section 5.15.

5.7. The Method of Marks

The method of marks (Section 2.4) can be usefully employed in discrete state stochastic processes. If $p_x(t)$ denotes the probability of x points of a point process lying in the half-open interval $(0, t]$, then the probability generating function

$$\phi(s, t) = \sum_{j=0}^{\infty} p_j(t) s^j$$

can be interpreted as the probability that there are no marked points in the interval $(0, t]$ when each point is marked independently with probability $1 - s$.

Poisson Process

To have no marked points in $(0, t + \Delta t]$ requires no marked points in $(0, t]$ and either no point in $(t, t + \Delta t)$ or else a point which is not marked:

$$\phi(s, t + \Delta t) = \phi(s, t)[1 - \lambda \Delta t + s\lambda \Delta t + o(\Delta t)],$$

which leads directly to the differential equation for the Poisson process (Section 5.3).

The method of marks can also be adapted to apply to Laplace transformations. Consider a renewal process with gap density $g(x)$, and suppose the marks are points of a completely independent Poisson process with parameter x. Then, beginning at the beginning of a gap (or at any other point), the probability of no mark before time t is e^{-st}, i.e., the tail of the negative exponential distribution. Thus the Laplace transform

$$\phi(s)=\int_0^\infty e^{-sx}g(x)dx$$

represents the probability that no mark occurs during the gap.

Thinning a Renewal Stream

Consider a process formed from a renewal process by removing each point with probability p. An application would be to the times between servicing machines, where the thinning corresponds to considering only those service times requiring a particular *type* of service. Let X be a renewal gap from one point of the thinned stream to the next point, whether removed or not, and let Y be a thinned gap from one point of the thinned stream to the next thinned stream point. Let X have density $g(x)$ and Laplace transform $\phi(s)$ and let Y have density $h(x)$ and Laplace transform $\psi(s)$. Superpose a Poisson stream with parameter s on the whole process. Then

$$\psi(s)=P(\text{no Poisson point in }Y). \qquad (24)$$

Condition this on whether or not the point τ which ends the gap X is retained in the thinned stream or not:

$P(\text{no Poisson point in }Y)= P(\text{no Poisson point in }Y|\tau\text{ retained}) P(\tau\text{ retained})$

$+P(\text{no Poisson point in }Y|\tau\text{ dropped}) P(\tau\text{ dropped})$

$= P(\text{no Poisson point in }X)(1-p)$

$+P(\text{no Poisson point in }X) P(\text{no Poisson point in }Y)p,$

which is equivalent to

$$\psi(s)=(1-p)\phi(s)+p\phi(s)\psi(s). \qquad (25)$$

Note in this derivation the key fact that if the point terminating the gap X is dropped, then the interval Y consists of independent intervals X and Y, the latter beginning with the dropped point. Thus

$$\psi(s)=\frac{(1-p)\phi(s)}{1-p\phi(s)}. \qquad (26)$$

Some special cases: (i) If $\phi=\psi$, then $p=0$. (ii) If ϕ is negative exponential with parameter λ, then ψ is also negative exponential with parameter

$(1-p)\lambda$. (iii) If $\phi(s)$ is Erlang with parameter k, it is easy to write down $\psi(s)$, but not easy to invert the Laplace transform. Even when $k=2$, the thinned process is complicated:

$$h(x)=\frac{1}{\lambda\pi^{1/2}}\left[\lambda^2(1-p)\right]e^{-\lambda x}\sin(\pi^{1/2}\lambda x),$$

according to *Tables of Integral Transforms*, Vol. 1.[†]

5.8. The Markov Infinitesimal Matrix

Consider the matrix defined in Section 5.5:

$$\Lambda=(\lambda_{xy})=\left(\left.\frac{dp_{xy}(t)}{dt}\right|_{t=0}\right). \qquad (27)$$

Several properties of this matrix shed light on the process. Since diagonal elements $p_{xx}(t)$ have the value $+1$ at $t=0$ and decline to zero at $t=\infty$ (by the property of negative-exponential sojourn), λ_{xx}, the slope at the origin, must always be negative. Similarly, since for nondiagonal elements $p_{xy}(t)$ is monotonically increasing from the value zero at the origin, nondiagonal λ_{xy} are always positive. Furthermore, each row sum in the matrix must be zero, since it is the derivative of $\Sigma_y p_{xy}(t)=1$.

Writing the Markov transition probabilities also in matrix form,

$$\mathbf{P}(t)=\left(p_{xy}(t)\right),$$

the forward and backward differential equations can be written in the matrix form

$$\mathbf{P}'(t)=\mathbf{P}(t)\Lambda \qquad \text{(forward equations)},$$

and

$$\mathbf{P}'(t)=\Lambda\mathbf{P}(t) \qquad \text{(backward equations)}. \qquad (28)$$

The (negative) diagonal elements in the infinitesimal matrix are exactly the (negative exponential) parameters of the sojourn times. Suppose a

[†]ERDELYI, A., ed. (1954), McGraw-Hill, New York, p. 229, Eq. (4).

sojourn time has length τ, with parameter β:

$$P(\tau < x) = 1 - e^{-\beta x}.$$

Then

$$P(\tau > \Delta t) = 1 - \beta \Delta t + o(\Delta t).$$

But $P(\tau > t)$, the probability that the sojourn time exceeds t, is the probability that a process in state x does not change to another state in time t:

$$p_{xx}(\Delta t) = 1 + \lambda_{xx} \Delta t + o(\Delta t),$$

and therefore, given that the system is in state x, the negative exponential parameter is $\beta = -\lambda_{xx}$.

Neglecting the sojourn times, the process behaves like a discrete time Markov chain, with the transition matrix having zeros on the diagonal and $-\lambda_{xy}/\lambda_{xx}$ elsewhere. The Poisson continuous time Markov infinitesimal matrix

$$\begin{pmatrix} -\lambda & \lambda & 0 & \cdots \\ 0 & -\lambda & \lambda & \cdots \\ 0 & 0 & -\lambda & \cdots \\ \vdots & \vdots & \vdots & \end{pmatrix}$$

thus corresponds to the discrete Markov chain matrix

$$\begin{pmatrix} 0 & 1 & 0 & 0 \cdots \\ 0 & 0 & 1 & 0 \cdots \\ 0 & 0 & 0 & 1 \cdots \\ \vdots & \vdots & \vdots & \vdots \end{pmatrix}.$$

However, it is not true that every discrete time Markov chain can be regarded as a continuous time Markov process simply by introducing negative exponential sojourn times, since there would be no provision for the case where a chain goes from state x to state x. Introducing two negative exponential sojourns with parameter λ_{xx} would destroy the Markov property of the process, since the sum of the two sojourns would be gamma distributed and would represent the time between transitions.

In this correspondence there is an interesting difference between the stationary vector $\{\pi_x\}$ of the chain and the equilibrium vector $\{p_x\}$ of the process.

As an illustration, consider the two-state continuous time Markov process of Section 5.4. The equilibrium probabilities have been shown to be

$$\left(\frac{\mu}{\lambda + \mu}, \frac{\lambda}{\lambda + \mu} \right).$$

Which two-state Markov chain does this correspond to? Since the transition probabilities were proportional to λ and μ, the first guess might be the chain with matrix

$$\begin{pmatrix} 1 - \lambda & \lambda \\ \mu & 1 - \mu \end{pmatrix},$$

which indeed does have the equilibrium probabilities given above. But it could be argued that the "right" discrete time analog would be a chain with matrix

$$\begin{pmatrix} 0 & 1 \\ 1 & 0 \end{pmatrix},$$

since there is no probability of a transition to the same state in the continuous case, and for this chain the equilibrium probabilities are $(\frac{1}{2}, \frac{1}{2})$. The problem of which chain corresponds to the Markov two-state process is in fact exactly the "stick problem" of Section 2.2, with the sticks being the sojourn times. The model leading to equiprobable equilibrium corresponds to "choosing a stick" (i.e., a state), while the other model is "length biased" according to sojourn length. In this light, it is natural that the continuous time model should agree with the length-biased discrete time model.

In Section 5.10, the relationship will be examined more systematically, and a formula connecting the two distributions will be established.

It is worth noting that the two distributions agree in this instance when $\lambda = \mu$, that is, when the point process is Poisson. This is true more generally and merely reflects that Poisson points are "at random," and so probabilities evaluated at these points are not length biased towards either state.

5.9. The Renewal Function

The notation of this section is consistent with that of Sections 5.2 and 5.3: Let $N(t)$ represent the number of renewals in the interval $(0, t]$, with gap density $g(x)$, distribution function $G(x)$, mean gap γ. Thus $N(t)$ is the random variable defining the counting distribution for the renewal process. The expected value of this random variable is a function of t and is called the *renewal function*:

$$M(t)=E(N(t))= \sum_{x=1}^{\infty} xP(N(t)=x). \qquad (29)$$

In order to express the renewal function in terms of the gap distribution, it is necessary to consider once again the fundamental relation between gaps and counts. The event that less than x renewal points occur in time t is exactly the event that the xth renewal point is beyond t:

$$N(t)<x \qquad \text{if and only if } \tau_x>t.$$

Thus

$$P(N(t)<x)=\int_{t}^{\infty} g^{x*}(u)du, \qquad (30)$$

because the convolution $g^{x*}(u)$ is the distribution of $\tau_x=\sigma_0+\sigma_1+\cdots+\sigma_{x-1}$, a sum of x independent, equidistributed random variables. Since the mean can be expressed as the sum of tails (cf. Problem 37, Chapter 1),

$$M(t)= \sum_{x=1}^{\infty} \int_{0}^{t} g^{x*}(u)du. \qquad (31)$$

This slightly complicated formula, involving as it does a sum, an integral, and a multiple convolution, can be somewhat simplified by taking the Laplace transform (with respect to t). Defining

$$\beta(s)=\int_{0}^{\infty} e^{-st}M(t)dt, \qquad \phi(s)=\int_{0}^{\infty} e^{-sx}g(x)dx$$

and using the various relationships obtained in Section 4.11 gives

$$\beta(s)=\frac{\phi(s)}{s[1-\phi(s)]} \qquad (32)$$

as the relation between the gap and the renewal function Laplace transforms. For some purposes, it is convenient to consider the derivative

$m(t)=dM/dt$, sometimes called the *renewal density* function—although it is not a density in the sense of a probability density. Since the Laplace transform

$$\alpha(s)=\int_0^\infty e^{-st}m(t)dt$$

of the renewal density is just $s\beta(s)$, a still simpler formula results:

$$\alpha(s)=\frac{\phi(s)}{1-\phi(s)}. \tag{33}$$

Equation (33) can be inverted into an integral equation,

$$m(t)=g(t)+\int_0^t g(u)m(t-u)du,$$

which is called the *integral equation of renewal theory*. It is possible to obtain this integral equation by direct probabilistic argument, giving an alternative derivation of Eqs. (32) and (33). To do so, it is necessary only to note that $m(t)$ can be interpreted as the mean number of renewals in a narrow interval near t:

$$m(t)=\lim_{\Delta t\to 0+}\frac{P(\geqslant 1 \text{ renewals in } (t,t+\Delta t))}{\Delta t}$$

$$=\sum_{x=1}^\infty g^{x*}(t).$$

Equation (33) can also be established by the method of marks. Suppose a Poisson process with parameter s is superposed on the renewal process. Then $\phi(s)$ represents the probability that a renewal point occurs before a Poisson point, and $\alpha(s)$ is the expected number of renewal points before the first Poisson point:

$$\phi(s)=\frac{\alpha(s)}{1+\alpha(s)}. \tag{34}$$

The relationship between the Laplace transform of the renewal density (or the renewal function) on the one hand and the Laplace transform of the gap distribution on the other shows that the renewal process is completely specified by the renewal function or the renewal density. This in turn shows the importance of the renewal function.

There are two important theorems relating to $N(t)$ and $M(t)$, the proofs of which are beyond the scope of this book, but which are intuitively clear:

Theorem 1 (Elementary Renewal Theorem)

$$\lim_{t \to \infty} \frac{M(t)}{t} = \frac{1}{\gamma}. \qquad (35a)$$

Theorem 2

$$\lim_{t \to \infty} \frac{N(t)}{t} = \frac{1}{\gamma}. \qquad (35b)$$

It should be noted, however, that the first of these theorems follows from the (unproved) $n=1$ generalization of Theorem 2, Section 4.11:

$$\lim_{t \to \infty} \frac{M(t)}{t} = \lim_{s \to 0} s^2 \beta(s) = \lim_{s \to 0} \frac{s\phi'(s)+\phi(s)}{-\phi(s)} = \frac{1}{\gamma}.$$

By L'Hôpital's rule, the first theorem implies that

$$\lim_{t \to \infty} m(t) = \frac{1}{\gamma},$$

which is entirely in accordance with the intuitive interpretation of the renewal density function $m(t)$ as the expected number of renewals in a small interval surrounding the point t.

5.10. The Gap Surrounding an Arbitrary Point

In the stick problem of Section 2.2, and especially in the discrete/continuous sojourn time problem of Section 5.8, it has been indicated that choosing an arbitrary point is quite different from choosing an arbitrary interval. In particular, the interval surrounding an arbitrary point is different in its stochastic aspects from a typical interval. This means that it would not be expected that the renewal distribution $g(x)$ would also be the distribution of the interval surrounding an arbitrary point.

Although the emphasis so far has been towards denying any paradox in this situation, it might be good at this stage to explain why the word paradox is sometimes used. Suppose the renewal process refers to light bulbs. Then, looking at the light bulb which is now burning—if one is justified in regarding "now" as an arbitrary point in time—is not equivalent to looking at a typical light bulb. The bias is in the direction of especially long-lived light bulbs, since "now" is more likely to occur during one of the long lives than during one of the short ones, just as in the case of the stick problem.

In this section the problem of a "time-typical" interval, that is, the gap surrounding an arbitrary point, will be more precisely formulated and solved. It should be understood, in the first place, that the solution must involve some kind of limiting procedure. This is because an "arbitrary" point in time can only be chosen over a finite domain (rectangular distribution). It will therefore be necessary to choose a point t, solve the problem, and then let $t \to \infty$.

Three random variables are of interest: $X(t)$, the time back to the most recent renewal; $Y(t)$, the time forward to the next renewal; and $Z(t)$, the time from the most recent renewal to the next renewal. Although $Z(t) = X(t) + Y(t)$, it should be noted that the distribution of $Z(t)$ cannot be obtained as the convolution of the separate distributions, since $X(t)$ and $Y(t)$ are not independent. Also, the ranges of $Y(t)$ and $Z(t)$ are $(0, \infty)$, while that of $X(t)$ is $(0, t)$.

Depending on the application, these random variables have been given various names, as shown in Table 5.1. Using this terminology, in the simple stick case of Section 2.2, it would be appropriate to say that $(\frac{1}{2}, \frac{1}{2})$ is the *gap distribution*, and $(\frac{1}{3}, \frac{2}{3})$ is the *spread distribution*.

Theorem 1. The density function for the distribution of spread is

$$
\begin{aligned}
g(x)\big[M(t) - M(t-x)\big], & \quad t > x, \\
g(x)\big[1 + M(t)\big], & \quad t \le x.
\end{aligned}
\tag{36}
$$

Table 5.1. Names of random variables

$X(t)$	$Y(t)$	$Z(t)$
Deficit	Excess	Spread
Age	Residual lifetime	
Backward recurrence time	Forward recurrence time	
	Excess life	

Proof. Let $C(x|t)=P(Z(t)<x)$, and let $c(x|t)=dC/dx$. Then

$$C(x|t)= \sum_{r=0}^{\infty} P(Z(t)<x, \tau_r<t<\tau_{r+1})$$

$$= \sum_{r=0}^{\infty} P(\tau_{r+1}-\tau_r<x, \tau_r<t<\tau_{r+1})$$

$$= \sum_{r=0}^{\infty} P(\sigma_r<x, t-\sigma_r<\tau_r<t).$$

Since σ_r and τ_r are independent, differentiation with respect to x yields, for $t>x$,

$$c(x|t)=g(x) \sum_{r=0}^{\infty} \int_{t-x}^{t} g^{r*}(u)\,du$$

$$=g(x)(M(t)-M(t-x)).$$

Similarly, when $t\le x$, differentiation yields

$$c(x|t)=g(x) \sum_{r=0}^{\infty} \int_{0}^{t} g^{r*}(u)\,du=g(x)[1+M(t)]. \qquad \square$$

Theorem 2. The density function for the distribution of age is

$$[1-G(x)][\delta(t-x)+m(t-x)], \tag{37}$$

where, using the convention of Section 4.10 for mixed discrete and continuous distributions, $\delta(t-x)=1$ when $t=x$, and $\delta(t-x)=0$ otherwise.

Proof. Let $A(x|t)=P(X(t)<x)$, with $a(x|t)=dA/dx$. As in the proof of Theorem 1, consider separately the cases $\tau_r<t<\tau_{r+1}$, $r=0,1,2,\ldots$. Since $t\ge X(t)$ always, the first interval is rather special. For $r=0$, $X(t)=t$, so that the contribution to the density is simply a deterministic component at the value $x=t$.

When t falls in the interval (τ_r, τ_{r+1}), the value of $X(t)$ is stochastic, namely, $X(t) = t - \tau_r$. Therefore, omitting for the moment the first term,

$$A(x|t) = P(X(t) < x)$$

$$= \sum_{r=1}^{\infty} P(X(t) < x, \tau_r < t < \tau_{r+1})$$

$$= \sum_{r=1}^{\infty} P(\tau_r > t - x, \sigma_r > t - \tau_r).$$

Differentiating and keeping in mind the independence of σ_r and τ_r, gives

$$a(x|t) = \sum_{r=1}^{\infty} g^{r*}(t-x)[1-G(x)]$$

$$= m(t-x)[1-G(x)]$$

or, with the discrete component with probability $1 - G(x)$ added,

$$a(x|t) = [\delta(t-x) + m(t-x)][1-G(x)], \qquad 0 < x < t. \qquad \square$$

Theorem 3. The density function for the distribution of excess is

$$g(t+x) + \int_0^t m(t-u)g(x+u)\,du. \qquad (38)$$

Proof. Let $B(x|t) = P(Y(t) < x)$, with $b(x|t) = dB/dx$. Then

$$B(x|t) = \sum_{r=0}^{\infty} P(Y(t) < x, \tau_r < t < \tau_{r+1})$$

$$= \sum_{r=0}^{\infty} P(t < \sigma_r + \tau_r < t + x, \tau_r < t)$$

$$= \int_0^t \sum_{r=0}^{\infty} P(t < \sigma_r + u < t + x) g^{r*}(u)\,du$$

$$= \int_0^t \sum_{r=0}^{\infty} \int_{t-u}^{t+x-u} g(v) g^{r*}(u)\,dv\,du.$$

Differentiating with respect to x gives

$$b(x|t) = \int_0^t \sum_{r=0}^{\infty} g(t+x-u) g^{r*}(u) \, du.$$

Since $g^{0*}(u) = \delta(u)$, the first term becomes $g(x+t)$ and the remaining terms can be written

$$\int_0^t g(u+x) m(t-u) \, du,$$

which gives the required result.   □

Theorem 4. In the limit as t approaches infinity, the densities of $X(t)$, $Y(t)$, and $Z(t)$ approach respectively

$$\frac{1}{\gamma}[1-G(x)], \qquad \frac{1}{\gamma}[1-G(x)], \qquad \frac{xg(x)}{\gamma}. \tag{39}$$

Proof. These results follow directly from the first three theorems and the limits for $m(t)$ and $M(t)$ given in section 5.9. The three limiting distributions will be denoted by $a(x)$, $b(x)$, and $c(x)$.   □

Theorem 5. The Laplace transforms of the distributions of $X(\infty)$, $Y(\infty)$, and $Z(\infty)$ are respectively

$$\frac{1-\xi(s)}{\gamma s}, \qquad \frac{1-\xi(s)}{\gamma s}, \qquad -\frac{\xi'(s)}{\gamma}, \tag{40}$$

where $\xi(s)$ is the Laplace transform of $g(x)$.

Some Definitions. A renewal process in which the first interval σ_0 has probability density $[1-G(x)]/\gamma$ while the other renewal intervals are independently $g(x)$ is called (for obvious reasons) an *equilibrium renewal process*, and a renewal process in which the first interval has probability density $h(x)$, not connected with the density $g(x)$ of the other intervals, is called a *generalized renewal process*. It is sometimes convenient to refer to the *ordinary renewal process* in contrast to the equilibrium and generalized process—for the ordinary process, all intervals are independently equidistributed, as treated so far in this chapter.

Theorem 6. For the Poisson process, with parameter λ (i.e., $\gamma = 1/\lambda$), the following values are obtained:

$$\text{(i)} \qquad M(t) = \lambda t,$$

$$\text{(ii)} \qquad m(t) = \lambda,$$

$$\text{(iii)} \qquad a(x|t) = e^{-\lambda x}\big[\delta(t-x) + \lambda\big],$$

$$\text{(iv)} \qquad b(x|t) = \lambda e^{-\lambda x},$$

$$\text{(v)} \qquad c(x|t) = \lambda^2 x e^{-\lambda x}, \qquad t > x,$$

$$\qquad\qquad\qquad = \lambda e^{-\lambda x}(1 + \lambda t), \qquad t \leq x,$$

$$\text{(vi)} \qquad a(x) = b(x) = g(x) = \lambda e^{-\lambda x},$$

$$\text{(vii)} \qquad c(x) = \lambda^2 x e^{-\lambda x} = g^{2*}(x).$$

(41)

Proof. This will be left as an exercise for the student. □

In connection with the "paradoxical" aspects of the problem, it is worth noting that the asymptotic mean spread in the Poisson case is $2/\lambda$, double the mean gap. Thus one might say that the light bulb now burning has double the lifetime of the average bulb.

It should also be noted that, naturally, $a(x|0) = \delta(x)$, $b(x|0) = c(x|0) = g(x)$.

5.11. Counting Distributions

A distribution which gives the probability of x events in time t is called a counting distribution; it gives the probability that a counting process is in state x at time t. Except for a process with constant aging, the counting distribution will depend on the location of the initial point. In Section 5.1 it was assumed that the interval began just after the origin and, furthermore, that an event occurred at the origin. This formulation is valid, therefore, for an interval of length t beginning at any event.

The general formulation, on the other hand, would relate to an interval beginning exactly T units after an event. Such a formulation could begin with considerations similar to those of Section 5.10 and would lead to a counting distribution dependent both on the initial and terminal point of the interval.

Instead of embarking on such a general formulation, another special counting distribution will be discussed, in which the initial point of the counting interval is an arbitrary point in time. The expression "an arbitrary point in time" will be considered to be a point for which the forward recurrence time is the asymptotic forward recurrence time; by Theorem 4 of Section 5.10, it has the distribution $[1 - G(x)]/\gamma$. The expressions *synchronous* and *asynchronous* are used to describe the two types of counting distributions: one beginning with a gap $g(x)$ to the first event, and the other beginning with a gap $[1 - G(x)]/\gamma$ to the first event. The following notation will be used. Length of counting period: t. For synchronous counting distribution, each gap has density $g(x)$; the probability of x counts in time t is $p_x(t)$, with distribution function $P_x(t)$, tail $Q_x(t)$, probability generating function $\pi(z, t)$,[†] and mean count M. For asynchronous counting, the first gap has density $[1 - G(x)]/\gamma$ and the remaining gaps have density $g(x)$. The same notation will be used for the various distributions, transforms, etc., but with a bar: $\bar{p}_x(t)$, $\bar{P}_x(t)$, $\bar{Q}_x(t)$, $\bar{\pi}(z, t)$, $\bar{M}$.

In both the synchronous and asynchronous cases, the generalization of the Poisson–gamma relationship (as given in Section 5.3) holds, so that the synchronous counting distribution satisfies

$$P_x(t) = \int_t^\infty g^{x*}(u)\, du, \tag{42}$$

leading to a Laplace transform (with variable s)

$$\mathcal{L} Q_x(t) = \frac{1}{s} \phi^x(s), \tag{43}$$

[†] In this section, and in many which follow, Laplace transforms (of time) occur together with probability generating functions. As an aid to memory, the variable z will be used for the probability generating functions, and s for the Laplace transforms. Thus the variables s, t correspond through the Laplace transform, and the variables x, z correspond through the probability generating function. In functional form, x is written as a subscript, and the other three in parentheses. When z occurs with s or t, it is written before the comma, with the s or t variable after the comma. Naturally, this system does *not* apply in cases where Laplace transforms are represented in terms of probability generating functions, such as will occur in Chapter 6 (for example, Section 6.6). In these latter cases, the Laplace transform is not with respect to a time variable, but with respect to a continuous density function, and the letter s is retained for both the probability generating function and the Laplace transform.

where $\phi(s)$ is the transform of $g(x)$. Therefore

$$\mathcal{L}\pi(z,t) = \frac{1}{s} + \frac{1}{s}\phi(s)z + \frac{1}{s}\phi^2(s)z^2 + \cdots$$

$$- \frac{1}{s}\phi(s) - \frac{1}{s}\phi^2(s) - \cdots$$

$$= \frac{1-\phi(s)}{s[1-z\phi(s)]},\tag{44}$$

so that the probability generating function is the inverse transform of the fraction on the right and the tails are inverse transforms of $\phi^x(s)/s$.

Then standard procedures, or else the formulas of Section 5.9, give the mean value

$$M(t) = \mathcal{L}^{-1} \frac{\phi(s)}{s[1-\phi(s)]}.\tag{45}$$

In the asynchronous case, the calculations are much the same, starting with

$$\bar{P}_x(t) = \int_t^\infty \frac{1}{\gamma}[1-G(u)] * g^{(x-1)*}(u)\, du\tag{46}$$

and

$$\mathcal{L}\bar{Q}_x(t) = \frac{\phi^{x-1}(s)[1-\phi(s)]}{\gamma s^2},\tag{47}$$

leading to

$$\bar{\pi}(z,t) = \mathcal{L}^{-1}\left(\frac{1}{s} + \frac{[1-\phi(s)](z-1)}{s^2[1-z\phi(s)]}\right),\tag{48}$$

with mean value

$$\bar{M}(t) = t/\gamma,\tag{49}$$

as would be intuitively expected.

These formulas are selected from many which relate the synchronous and asynchronous counting distributions to the gap distribution and to one another. Others are given in the problem list for the chapter.

In the terminology of renewal theory, $\overline{M}(t)$ would be called the renewal function for an equilibrium renewal process, and $M(t)$ the renewal function for an ordinary renewal process.

The emphasis placed in this section on two particular types of counting procedures should not obscure the fact that the *general* renewal counting distributions, based on an arbitrary initial gap, are also of importance. One such process will be discussed in connection with particle counters in which the first gap has transform $\lambda/(\lambda+s)$ and the subsequent gaps have transform $\lambda e^{-\lambda\Delta}/(\lambda+s)$. These two transforms do not satisfy the relationship between gap and excess given by Theorem 5 of Section 5.10, on which the definition of asynchronous counting is based.

5.12. The Erlang Process

The Erlang process is defined by gamma-distributed gaps in a renewal process. It is a natural generalization of the Poisson process, since the gamma-distributed gaps become negative exponential when $k=1$. The Erlang process has two fundamental applications: First, it represents a process in which the events occur after k "stages," and second, it can be interpreted as a process where $k-1$ consecutive Poisson events are deleted.

The first application is exemplified by repair to a machine, in case the repair consists of k sub-repairs, or by service if the service consists of k sub-services, provided the sub-repairs or sub-services are such that the time taken to perform then is negative exponentially distributed. In the second application, the Erlang process would be applicable to cases where it is necessary to accumulate k Poisson events before a recorded event would take place. For example, k randomly occurring shocks might be required for a industrial item to be broken.

The Laplace transform of the Erlang gap distribution can be written

$$\phi(s)=\left(\frac{\lambda}{\lambda+s}\right)^{k},$$

so that the Laplace transform of the renewal density $m(t)$ is

$$\alpha(s)=\frac{\lambda^{k}}{(\lambda+s)^{k}-\lambda^{k}}. \tag{50}$$

Inversion of this transform is usually not easy, and for large values of k it is even necessary to use numerical methods. For $k=2$, however, there is a simple solution:

$$\alpha(s)=\lambda^2 s^{-1}(2\lambda+s)^{-1}.$$

Expanding into partial fractions or using the theorem of Section 4.12, the student will be able to invert this Laplace transformation to obtain

$$m(t)=\tfrac{1}{2}\lambda-\tfrac{1}{2}\lambda e^{-2\lambda t}. \tag{51}$$

From this result, it is easy to see by integration that the renewal function for the two-stage Erlang process is

$$M(t)=\tfrac{1}{2}\lambda t-\tfrac{1}{4}+\tfrac{1}{4}e^{-2\lambda t}. \tag{52}$$

The synchronous and asynchronous counting distributions are also fairly easy to write for general k and can, in fact, be represented conveniently in terms of the Poisson probabilities.

Synchronous Counting (Ordinary Renewal Process)

The general formulas reduce to the following special cases for $\phi(s)=\lambda^k/(\lambda+s)^k$:

$$\mathcal{L}\pi(z,t)=\frac{(\lambda+s)^k-\lambda^k}{s\left[(\lambda+s)^k-z\lambda^k\right]}, \tag{53}$$

$$P_x(t)=\frac{\Gamma(xk,\lambda t)}{\Gamma(xk)}. \tag{54}$$

Let

$$B_x=\frac{e^{-\lambda t}(\lambda t)^x}{x!}.$$

Then

$$p_x(t)=\sum_{j=xk}^{(x+1)k-1} B_j, \qquad x=0,1,2,\ldots. \tag{55}$$

Thus the probability of no events in time t is represented as the sum of the first k Poisson terms with parameter λt, the probability of one event is the sum of the next k Poisson terms, and so forth. The probabilities can therefore also be written in the form

$$p_x(t) = \sum_{j=0}^{k-1} \frac{e^{-\lambda t}(\lambda t)^{xk+j}}{(xk+j)!},$$

and the mean written

$$M(t) = \sum_{j=1}^{\infty} \frac{\gamma(kj, \lambda t)}{\Gamma(kj)}. \tag{56}$$

Asynchronous Counting (Equilibrium Renewal Process)

Define

$$A_x = \frac{\gamma(x, \lambda t)}{\Gamma(x)}$$

and begin with Eq. (i) of Problem 18 (Section 5.15). Then

$$\bar{Q}_x(t) = \frac{1}{\gamma} \sum_{j=(x-1)k}^{xk-1} \int_0^t B_j$$

$$= \frac{1}{k} \sum_{j=(x-1)k+1}^{xk} A_j, \tag{57}$$

from which the student can obtain $\bar{p}_x(t)$ and the other quantities characterizing the distribution.

5.13. Displaced Gaps

In several applications, it is important to consider a gap distribution which is identically zero over some range $(0, \Delta)$. For example, a particle counter may have a *dead time* of length Δ during which additional registrations are impossible. Also, in the theory of traffic flow, it is often important to take into consideration a fixed *car length* Δ, so that car arrivals are never nearer than a car length.

Consider a gap distribution $g_\Delta(x)$ which is based on a known gap distribution $g(x)$ by displacement:

$$g_\Delta(x)=g(x-\Delta), \qquad \Delta<x<\infty. \tag{58}$$

As a matter of notation, the Laplace transform will also be denoted by a subscript,

$$\phi_\Delta(s)=e^{-s\Delta}\phi(s), \tag{59}$$

and the counting distributions by an upper delta: $p_x^\Delta(t),\,\bar{p}_x^\Delta(t),\,P_x^\Delta(t),\,\bar{P}_x^\Delta(t),$ $Q_x^\Delta(t),\bar{Q}_x^\Delta(t)$.

Synchronous Counting

By Section 4.11,

$$Q_x^\Delta(t)=\mathcal{L}^{-1}\frac{\phi_\Delta^{x+1}(s)}{s}$$

$$=\mathcal{L}^{-1}e^{-s\Delta(x+1)}\frac{\phi^{x+1}(s)}{s}. \tag{60}$$

The Laplace inverse of $\phi^{x+1}(s)/s$ is simply $Q_x(t)$, and powers of s have the effect of differentiating the inverse, since $Q_x(0)=0$. Therefore

$$Q_x^\Delta(t)=\mathcal{L}^{-1}\frac{\phi^{x+1}(s)}{s}\sum_{j=0}^{\infty}\frac{[-s\Delta(x+1)]^j}{j!}$$

$$=\sum_{j=0}^{\infty}\frac{[-\Delta(x+1)]^j}{j!}\frac{d^jQ_x(t)}{dt^j}, \tag{61}$$

which can be expressed, using Taylor's theorem, as

$$Q_x^\Delta(t)=Q_x[t-\Delta(x+1)]. \tag{62}$$

This is a rather curious distribution, in that the argument on the right side becomes negative when $t<\Delta(x+1)$. In fact, since $Q_x^\Delta(t)$ represents the probability of more than x events beginning just after an event, such a count will be impossible when $t<\Delta(x+1)$. Therefore the result should be stated in full form:

$$Q_x^\Delta(t) = Q_x[t - \Delta(x+1)], \qquad t > \Delta(x+1),$$

$$Q_x^\Delta(t) = 0, \qquad\qquad\qquad t \le \Delta(x+1). \tag{63}$$

The Displaced Negative Exponential

If $g(x) = \lambda e^{-\lambda x}$, then the displaced distribution can be equally well regarded as *truncated*, by the fact that the negative exponential distribution does not exhibit aging. Let the kth Poisson probability with parameter $\lambda(t - j\Delta)$ be denoted by B_k^j:

$$B_k^j = (1/k!) e^{-\lambda(t - j\Delta)} [\lambda(t - j\Delta)]^k. \tag{64}$$

Then the exact probabilities can be obtained from the values for the tails as follows:

$$p_0^\Delta(t) = \begin{cases} B_1^1, & t > \Delta, \\ 1, & t \le \Delta, \end{cases}$$

$$p_1^\Delta(t) = \begin{cases} -B_0^1 + B_0^2 + B_1^2, & t > 2\Delta, \\ 1 - B_0^1, & \Delta < t \le 2\Delta, \\ 0, & 0 < t \le \Delta, \end{cases}$$

$$p_2^\Delta(t) = \begin{cases} -B_0^2 - B_1^2 + B_0^3 + B_1^3 + B_2^3, & t > 3\Delta, \\ 1 - B_0^2 - B_1^2, & 2\Delta < t \le 3\Delta, \\ 0, & t \le 2\Delta, \end{cases}$$

and, in general,

$$p_x^\Delta(t) = \begin{cases} \displaystyle\sum_{j=0}^{x} B_j^{x+1} - \sum_{j=0}^{x-1} B_j^x, & t > (x+1)\Delta, \\ 1 - \displaystyle\sum_{j=0}^{x-1} B_j^x, & x\Delta < t \le (x+1)\Delta, \\ 0, & t \le x\Delta. \end{cases}$$

The Poisson sums can be replaced by incomplete gamma functions according to Section 4.5, giving the formulas

$$
p_x^\Delta(t) = \begin{cases} \dfrac{\Gamma\{x+1, \lambda[t-(x+1)\Delta]\}}{\Gamma(x+1)} - \dfrac{\Gamma[x, \lambda(t-x\Delta)]}{\Gamma(x)}, & t > (x+1)\Delta, \\[3mm] 1 - \dfrac{\Gamma[x, \lambda(t-x\Delta)]}{\Gamma(x)}, & x\Delta < t \le (x+1)\Delta, \\[3mm] 0, & t \le x\Delta. \end{cases}
$$

(65)

5.14. Divergent Birth Processes

This section contains a treatment of a type of process in which there can be an infinite number of transitions in a finite time, just as a perfect ball might bounce an infinite number of times in a finite time. The idea is simply that the gaps converge. It is clear that this could never happen in a renewal process, since the gaps are equidistributed. However, the possibility exists for Markov processes; although the gaps are all negative exponential, the parameter in the negative exponential distribution could vary in some systematic way which would lead to an infinite number of transitions in a finite interval.

Such processes are called *divergent*; the counting distributions do not sum to unity since the range contains, besides the non-negative integers with probabilities $p_x(t)$, $x = 0, 1, 2, \ldots$, a special point (designated by ω), which indicates an infinite number of transitions in time t. Thus

$$
\sum_{j=0}^{\infty} p_j(t) = 1 - p_\omega(t)
$$

would merely mean that for normalization to unity, the special value must also be taken into account.

An important example of a divergent process is the birth process treated in Section 5.5, where the coefficients λ_x are assumed to increase more rapidly than the simple linear expression leading to the Yule process.

Let Z be the (finite or infinite) time required for an infinite number of transitions. Then the probability of a divergent process is the probability that Z is finite, that is $P(Z<t)$; in other words, the distribution function for Z represents the probability that an infinite number of transitions have taken place before time t. The density function for Z is

$$\left(\lambda_1 e^{-\lambda_1 t}\right) * \left(\lambda_2 e^{-\lambda_2 t}\right) * \cdots$$

and the Laplace transform of this quantity is

$$\left(\frac{\lambda_1}{s+\lambda_1}\right) \times \left(\frac{\lambda_2}{s+\lambda_2}\right) \times \cdots.$$

To obtain the Laplace transform of the distribution function it is necessary to multiply by $1/s$. Therefore, if

$$\beta(s) = \int_0^\infty e^{-su} P(X<u)\, du,$$

$\beta(s)$ can be written

$$\beta(s) = \frac{1}{s} \prod_{j=1}^\infty \left(1 + \frac{s}{\lambda_j}\right)^{-1}. \tag{66}$$

When the series converges, so will the infinite product, and this convergence is the necessary condition that $p_\omega(t)$ be nonzero.

There is a theorem of algebra[†] to the effect that a necessary and sufficient condition for the convergence of the infinite product (66) is the convergence of the series $\Sigma \lambda_j^{-1}$. Therefore the process will be divergent when

$$\sum_{j=1}^\infty \frac{1}{\lambda_j} < \infty. \tag{67}$$

When this convergence occurs, it is possible to write $\beta(s)$ as a series by using partial fractions:

$$\beta(s) = \frac{k_0}{s} + \sum_{j=1}^\infty \frac{k_j}{s+\lambda_j}, \tag{68}$$

where the k_j are constants to be determined. The inverse Laplace transform

---

[†]See, for example, BROMWICH, THOMAS JOHN I'ANSON (1908), *An Introduction to the Theory of Infinite Series*, Macmillan & Co., London. The proof of the theorem consists essentially of a close examination of the logarithm of the infinite product.

is now easy to write down:

$$p_\omega(t)=k_0+\sum_{j=1}^{\infty}k_je^{-t\lambda_j}.$$

To evaluate the constants k_j, the standard method of partial fractions is used. One denominator at a time is multiplied by the equation and then this quantity is equated to zero. Thus

$$k_0=1$$

and, for $j>0$,

$$k_j=\lim_{s\to-\lambda_j}\,(s+\lambda_j)\beta(s).\tag{69}$$

The problem of finding $p_\omega(t)$ thus depends on evaluating these limits, which in some cases can be rather difficult.

Example 1. $\lambda_j=j^2$, which guarantees the convergence of $\Sigma\lambda_j^{-1}$:

$$k_j=\lim_{s\to-j^2}\frac{s+j^2}{s(1+s)(1+\frac{1}{4}s)\cdots}$$

$$=-\left[(1-j^2)(1-\tfrac{1}{4}j^2)\cdots\left(1-\frac{j^2}{(j-1)^2}\right)\left(1-\frac{j^2}{(j+1)^2}\right)\cdots\right]^{-1}$$

$$=-2j\frac{1\cdot4\cdot9\cdots(j-1)^2(j+1)^2\cdots}{[(1-j)(2-j)\cdots(-1)1\cdot2\cdot3\cdots][(1+j)(2+j)\cdots]}$$

$$=2(-1)^j.$$

Therefore the probability of an infinite number of events before a finite time t is given by

$$p_\omega(t)=1+2\sum_{j=1}^{\infty}(-1)^je^{-tj^2}.$$

Example 2. $\lambda_j = \lambda^j$, similarly guaranteeing the convergence of $\Sigma \lambda_j^{-1}$ for $\lambda > 1$. Then

$$k_n = -\frac{1}{\lambda^n} \frac{\lambda}{\lambda - \lambda^n} \frac{\lambda^2}{\lambda^2 - \lambda^n} \cdots \frac{\lambda^{n-1}}{\lambda^{n-1} - \lambda^n} \Omega,$$

where

$$\Omega = \prod_{j=1}^{\infty} (1 - \lambda^j)^{-1}$$

and is thus independent of n. The ratio of the consecutive k_n therefore satisfies the difference equation

$$\frac{k_n}{k_{n+1}} = -\lambda(\lambda^n - 1),$$

with the initial condition

$$k_1 = -\Omega/\lambda.$$

It is easy to see that the solution of the difference equation is

$$k_n = \frac{(-1)^n \Omega}{\lambda^n (\lambda^{n-1} - 1)(\lambda^{n-2} - 1) \cdots (\lambda - 1)}. \tag{70}$$

In both examples it would be necessary to approximate the values of $p_\omega(t)$. In the second example, the following approximate formula[†] for Ω is useful:

$$\Omega^{-1} = \sum_{j=-\infty}^{\infty} (-1)^j (1/\lambda)^{j(3j+1)/2}. \tag{71}$$

Recommendations for Further Study

For an intermediate-level treatment of stochastic processes, both discrete and continuous, see Cox and Miller (1965) and Çinlar (1975). A self-contained book on renewal processes is that by Cox (1962). Some

[†]This rapidly converging series is given in ERDELYI, A. (1955), *Higher Transcendental Functions*, Vol. 3, McGraw-Hill, New York, p. 177.

important processes not treated in the present book are *Markov renewal processes* (see Çinlar, Chapter 10 and Ross, Chapter 7), *martingales* (see Karlin and Taylor, Chapter 6), and *processes based on the normal distribution* (*Gaussian, Wiener, etc.*) [see Hoel, Port, and Stone (1972) and Cox and Miller (1965)].

ÇINLAR, ERHAN (1975), *Introduction to Stochastic Processes*, Prentice-Hall, Englewood Cliffs, New Jersey.

COX, D. R. (1962), *Renewal Theory*, John Wiley and Sons, New York.

COX, D. R., and MILLER, H. D. (1965), *The Theory of Stochastic Processes*, Methuen, London.

HOEL, PAUL G., PORT, SIDNEY C., and STONE, CHARLES J. (1972), *Introduction to Stochastic Processes*, Houghton Mifflin, Boston.

KARLIN, SAMUEL, and TAYLOR, HOWARD M., (1975), *A First Course in Stochastic Processes*, Academic Press, New York.

ROSS, SHELDON M. (1972), *Introduction to Probability Models*, Academic Press, New York.

5.15. Problems[†]

1. Using the methods of Section 5.10, find the joint distribution of $X(t)$ and $Y(t)$, namely, density function

$$[\delta(t-x)+M(t-x)]g(x+y), \qquad 0<x<t, 0<y<\infty.$$

Use this result to prove Theorems 2 and 3 by direct integration and Theorem 1 by integration over the domain where $x+y=$ constant.

2. Consider a pure birth process in which the infinitesimal probabilities of a birth in Δt are $\lambda\Delta t+o(\Delta t)$ when there are an odd number alive at the beginning of the interval, and $\mu\Delta t+o(\Delta t)$ when there are an even number alive at the beginning of the interval. Define

$$A(t)=\sum_{j \text{ odd}} p_j(t), \qquad B(t)=\sum_{j \text{ even}} p_j(t).$$

Obtain the differential equations for $A(t)$ and $B(t)$ and solve.

3. The lifetime of a piece of machinery defines a renewal process with the usual notation: g, ϕ, etc. The machine is installed and working at $t=0$, and inspected

---

[†]Problem 6 is taken from Karlin and Taylor (1975), and Problems 9 and 10 are taken from Ross (1972), with kind permission of the publishers.

at $t=1,2,3,\ldots$. When it is first found to be broken, a new machine is installed and the process repeated. Let

$$Y_n = \begin{cases} 0 & \text{if the machine is broken at the } n\text{th inspection} \\ 1 & \text{if the machine is working at the } n\text{th inspection.} \end{cases}$$

$$Z_n = Y_1 + Y_2 + \cdots + Y_n,$$

that is, the (birthday) age of the machine, and

$$Z = \lim Z_j,$$

that is, the birthday age at which the machine breaks. (i) Show that Z_n is a Markov chain. (ii) Find the transition matrix in terms of g, G, ϕ, etc. (iii) Show that $P(Z=x)=G(x+1)-G(x)$.

4. In a Poisson process with parameter λ, suppose that N events occur in time t. Find the density function for X, the time to the nth event $(n<N)$.

   Ans. X/t is beta distributed.

5. Two machines are working, each with negative exponential lifetimes and parameter λ. When one fails, a replacement is furnished, the time to replace being also negative exponential with parameter μ. Let $X(t)$ be the number of working machines at time t, state space $(0,1,2)$. (i) Find the infinitesimal transition matrix. (ii) Find the forward equations. (iii) Solve the forward equations to find the transition probabilities $p_{xy}(t)$.

6. Consider a Yule process with parameter λ and initial state $N=1$. Suppose the probability of death of the original ancestor in time Δt is $\beta \Delta t + o(\Delta t)$, given that he is living at time t. Find the distribution of the number x of offspring from the single ancestor and his descendants at the time of his death.

   Ans. $(\beta/\lambda)B(\beta/\lambda+1, x+1)$

7. Suppose N identical balls are distributed into two boxes. A ball in box A (box B) remains there for a negative exponential time parameter λ (parameter μ) and goes to the other box. The balls act independently. Let $X(t)$ denote the number of balls in box A at time t. Then $X(t)$ is a birth-and-death process defined over $0, 1, \ldots, N$. (i) Find the birth and death rates. (ii) Find $p_{xN}(t)$. (iii) Find $E(X(t))$.

8. In a birth-and-death process, $\lambda_x = (1+x)^{-1}$ and $\mu_x = \mu$. Write down the forward equations.

9. A machine in use is replaced either when it fails or when it reaches an age of T. If the lifetimes of successive machines are independent with a common density $f(x)$ and cumulative form $F(x)$, show that the long-run rate at which machines are replaced is

$$\left(\int_0^T xf(x)\,dx + T(1-F(T)) \right)^{-1}.$$

10. It might appear that the finiteness of $m(t)$ would follow from the fact that $N(t)$ is finite. However, such reasoning is not valid, as the following example shows. Let Y be a random variable with $P(Y=2^n)=(\frac{1}{2})^n$, $n=1,2,3,\dots$. Then $P(Y<\infty)=1$, but $E(Y)=\infty$, as the student should show.

11. Consider two independent Poisson processes. Show that the distribution of the number of events in one process which fall between two consecutive events in the other process is geometric.

12. Consider a Poisson process with parameter λ. Find the distribution of the number of Poisson points which occur in an independent interval T which is gamma distributed with parameters μ and k.

13. A machine breaks when N shocks have been received. If the shocks occur at times which form a Poisson process with parameter λ, find the density function for the lifetime of the machine.

14. Let $X(t)$ be a Markov process with state space $(0,1)$ and transition probabilities

$$p_{xy}(t)=\tfrac{1}{2}\left[1+(-1)^{x+y}e^{-2t}\right], \qquad x,y=0,1.$$

Show that the Chapman–Kolmogorov equations are satisfied.

15. In a pure birth process with linear birth rate $\lambda_x=\lambda x$, let the initial number present be geometrically distributed, i.e., $P(N(0)=x)=(1-\rho)\rho^{x-1}, x=1,2,3,\dots$. Find $E(N(t))$.

16. Find the means of the distributions $a(x|t)$, $b(x|t)$, and $c(x|t)$ as functions of t.

17. Show that in the asymptotic case the gap variance can be expressed in terms of the mean spread and the harmonic mean spread. Note: the *harmonic mean* of a random variable is defined to be $[E(1/X)]^{-1}$.

18. In the notation of Section 5.11, prove the following formulas:

(i)
$$\bar{Q}_x(t)=\frac{1}{\gamma}\int_0^t p_{x-1}(u)\,du,$$

(ii)
$$\frac{\bar{\pi}(z;t)-1}{z-1}=\frac{1}{\gamma}\int_0^t \pi(z;u)\,du,$$

(iii)
$$\frac{d\bar{\pi}}{d\tau}=(z-1)\frac{\pi(z;t)}{\gamma}.$$

19. For the Erlang process with $k=2$, obtain the renewal density function from the renewal integral equation.

20. For the Erlang process with $k=2$, find the exact and asymptotic distributions of spread, excess and deficit.

21. Show that the two examples of divergent birth processes given in Section 5.14 lead respectively to the following equations for the probability generating function of the "finite" probabilities $p_x(t)$, $x = 0, 1, 2, \ldots$:
(i) For $\lambda_j = j^2$,

$$\frac{\partial \phi(z, t)}{\partial t} = -\lambda(1-z)\left(z^2 \frac{\partial^2 \phi}{\partial z^2} + z \frac{\partial \phi}{\partial z}\right).$$

(ii) For $\lambda_j = \lambda^j$,

$$\frac{\partial \phi(z, t)}{\partial t} = -(1-z)\phi(\lambda z, t).$$

22. Referring to Theorem 4 of Section 5.10, let the limiting moments of the distribution of $Y(t)$ be

$$\nu_n = \int_0^\infty x^n \frac{1 - G(x)}{\gamma} dx,$$

and let the corresponding quantities for the gap distribution be μ_n. Show that

$$n\mu_1 \nu_{n-1} = \mu_n.$$

23. Show that for displaced negative-exponential gaps (Section 5.13) the asymptotic distribution of excess (or deficit) given in Section 5.10 has Laplace transform

$$\frac{\lambda(\lambda+s) - \lambda^2 e^{-s\Delta}}{s(1+\lambda\Delta)(\lambda+s)},$$

and find the mean excess (or deficit) $E(X(\infty))$.

24. In the Erlang $k = 2$ case, replace the partial fraction argument of Section 5.12 used to obtain $m(t)$ from $\alpha(s)$ by the formula for Laplace inversion given in Section 4.12.

25. In Section 5.12, express $\bar{Q}_x(t)$ in terms of sums of B_x, without integration.

26. Find the means and variances of the distributions defined in Theorems 4 and 5 of Section 5.10.

27. Obtain the *backward equations*

$$p'_{xy}(t) = \sum_{j \neq x} \lambda_{xj} p_{jy}(t) - p_{xy}(t) \sum_{j \neq x} \lambda_{xj}$$

from

$$p_{xy}(t + \Delta t) = \sum_j p_{xj}(\Delta t) p_{jy}(t).$$

28. Find the failure rate for an Erlang process.

29. In thinning an Erlang process, find the limiting form of the thinned process as the variance of the gap distribution approaches zero.

30. In the discussion of renewal theory, it has been assumed that the lifetimes are finite. What modifications would need to be made if $G(\infty)<1$?

31. Consider a deterministic process $g(x)=\delta(x-\Delta)$. Define N so that the counting interval t satisfies

$$N\Delta\leq t\leq(N+1)\Delta,$$

that is, N is the maximum number of gaps that will fit into the counting period: $N=0,1,2,\dots$. Find the probability that there are x points in t, assuming that t is equally likely to begin anywhere in a gap. Show that the probability generating function of the counting distribution is

$$\frac{z^N}{\Delta}(\Delta-T+N\Delta+zT-zN\Delta).$$

Find the mean count.

32. In a renewal process, the lifetime (gap) distribution is *discrete* geometric with parameter ρ over the positive integers. Show that $P(\sigma_n=x)$ is negative binomial. Write down the probability generating function $\pi(z,t)$ of the synchronous counting distribution and find the mean $M(t)$.

33. Find the survivor function (Section 5.2) for (i) the Erlang process, (ii) the deterministic process, (iii) the process in Problem 32.

34. Show that the mean population size in a Yule process increases exponentially.

35. Discuss a two-state process with Erlang transition intervals. Obtain regular oscillation as a special limiting case.

36. In the nonequilibrium linear birth-and-death process specified by Eq. (23), show that the probability generating function $\phi(s,t)=p_j(t)s^j$ satisfies the partial differential equation

$$\frac{\partial\phi}{\partial t}=\left[\mu-(\lambda+\mu)s+\lambda s^2\right]\frac{\partial\phi}{\partial s}.$$

37. By solving the equation of Problem 36, or directly from Eq. (23), show that the nonequilibrium linear birth-and-death probabilities for $\lambda=\mu$ are given by

$$p_0(t)=\frac{\lambda t}{1+\lambda t},\qquad p_x(t)=\frac{(\lambda t)^{x-1}}{(1+\lambda t)^{x+1}},\qquad x=1,2,3,\dots.$$

38. Consider a linear birth-and-death process with a finite number of states n, so that $\lambda_x=(n-x)\lambda$, $\mu_x=\mu x$, $0\leq x\leq n$. Write down the basic differential equations of the process, and show that the equilibrium probabilities form a binomial distribution.

6

The Theory of Queues

6.1. Introduction and Classification

Queueing theory is a generic term for mathematical structures inspired by and descriptive of service systems with random features such as random delay, randomly arriving customers, etc.[†] Such systems can be classified in two ways: according to the structure and postulates which characterize the operation, on the one hand, and according to the random variable of interest, on the other. Table 6.1. with some of the important random variables, together with associated notation, is given on p. 228.

Structural classification of queues is by no means easy. There is not any single method which provides for all variations in systems; these variations quite often arise from the consideration of special applications. There is, however, a generally accepted rough classification due to Kendall,[‡] which describes in compact notation several of the more important properties of a system.

The Kendall symbol consists of three parts: The first gives information about the *arrival* stream of new customers, the second gives corresponding information about the *service* process, while the third element is simply an integer specifying the number of servers in the system. In each of the first two elements, the following abbreviations are used: M denotes a Poisson process, E_k denotes an Erlang process with parameter k, G ("general")

[†] The expression "stochastic service systems," due to John Riordan, expresses the idea of a queue quite well.
[‡] David George Kendall, English mathematician, 1918–

denotes a renewal process, D ("deterministic") denotes a point process with all intervals equal.

These symbols have slightly different interpretations when they appear as the first element (arrival) than when they appear as the second element (service). Thus an $M/M/1$ queue—the simplest type classified under the Kendall symbol—denotes a queue with arrivals joining the queue at instants which form a Poisson process. In some cases it is desirable to specify the Poisson parameter and write, for example, M_λ. For service, the letter M refers to the fact that the service times are negative exponentially distributed with the same parameter. The number "1" means simply that there is a single server. Thus the queue $M_\lambda/M_\mu/1$ has been treated already in Section 5.6, and the equilibrium distribution of the number in the system has been found to be geometric. In the next section, the nonequilibrium solution to this simple queueing problem will be found.

The various symbols, applied either to arrival or to service times, have naturally a variety of meanings in applied problems. As an arrival pattern, "D" might be called "scheduled" arrivals; applied to service, it might be considered "automated" service. Erlang arrivals might be nothing more than a two-parameter generalization of Poisson arrivals, or it might denote arrivals where each kth Poisson arrival is retained and the remainder "lost." Applied to service, the Erlang pattern could represent service consisting of k negative exponential stages. General (i.e., renewal) service would require only that the service periods be independently equidistributed.

A little reflection on a few of the applications of queueing theory will suggest various ramifications in the structure not covered by the Kendall symbol: aircraft waiting to land (where a fuel consumption constraint may induce priorities in the queue), customers waiting for taxis combined with taxis awaiting customers (where two queues confront one another), customers waiting in a food market (where there are different types of commodities in the basket), cars waiting for a traffic light (where the light may be vehicle actuated), customers awaiting an essentially unimportant service (where they may go away if the queue is too long—this is called *balking*), industrial machines awaiting repair (where repairs required may differ substantially), queues formed in networks, in series, or in parallel, where one queue may feed into several others—each of these examples suggests special requirements in the model. However, it is also true that in each case the Kendall symbol may be useful in describing some aspect of the system.

In this chapter no attempt is made to treat all of the most important varieties of queues; where a method is shown once in a particular context, it will not be repeated. The chapter is therefore a survey of methods useful in queueing theory rather than of different types of queues. With technique assimilated, the student can then investigate new structures, abstract from their real behavior an appropriate mathematical model of their ideal behavior, discover random variables of interest, and very likely have an approximate idea of the mathematical techniques appropriate to the problem.

In pursuing this goal, the chapter begins by considering several random variables of particular interest in relatively simple queues: Sections 6.2–6.7 deal with the total number in the system at time t, Sections 6.8 and 6.9 deal with the waiting time in the queue, and Sections 6.11–6.14 deal with the duration of the busy period. The remaining sections treat some special situations of general interest and importance.

The notation used in this chapter is systematic and is summarized for convenient reference in Table 6.1.

6.2. The $M_\lambda/M_\mu/1$ Queue: General Solution

This section begins with the equations for the queue as a birth-and-death process, specializing those obtained in Section 5.6:

$$p_x'(t) = -(\lambda+\mu)p_x(t) + \lambda p_{x-1}(t) + \mu p_{x+1}(t), \qquad x=1,2,3,\ldots,$$
$$p_0'(t) = -\lambda p_0(t) + \mu p_1(t), \tag{1}$$

where $p_x(t) = P(X(t)=x)$, $X(t)$ denotes the number of customers in the system, either being served or waiting for service, and where the instants when arrivals join the queue form a Poisson process with parameter λ and where the service times are negative-exponentially distributed with mean service time $1/\mu$. Multiplying the jth equation by z^{j+1}, it is not difficult to show that the probability generating function[†]

$$\phi(z,t) = \sum_{j=0}^{\infty} p_j(t)z^j$$

satisfies the differential equation

$$z\frac{d\phi}{dt} = (1-z)[(\mu-\lambda z)\phi(z,t) - \mu p_0(t)] \tag{2}$$

[†]See footnote in Section 5.11.

Table 6.1. Queueing notation (This system does not apply in Chapter 5)

Name	Random variable	Density	Probabilities	Distribution function	Laplace transform	Probability generating function	Mean	Variance
Interarrival time	U	$d(x)$			$\delta(s)$		$1/\lambda$	
Service time		$b(x)$		$B(x)$	$\beta(s)$		$1/\mu$	v
Number of departures in an interarrival period[a]			r_x			$\eta(s)$		
Number of arrivals in a service period	N		q_x			$\theta(s)$		
Number in the system	X		p_x			$\phi(s)$		
Number queueing	Y							
Queueing time	V	$a(x)$[b]		$A(x)$	$\alpha(s)$			
Waiting time	W	$c(x)$		$C(x)$	$\gamma(s)$			
Residual service time	$\overline{U}$				$\overline{\beta}(s)$			
Discrete busy period	K		h_x			$\kappa(s)$		
Continuous busy period	L	$f(x)$			$\sigma(s)$			
Balking	X'		g_x	G_x				

[a]Assuming the queue is never empty.
[b]The continuous portion.

for $0 \leq z \leq 1$. Let the original number in the system be n, so that $p_n(0) = 1$, $\phi(z,0) = z^n$. It is convenient to reduce the differential equation to an algebraic equation by using the Laplace transform:

$$\psi(z,s) = \int_0^\infty \phi(z,t)e^{-ts}dt,$$

$$\pi(s) = \int_0^\infty p_0(t)e^{-ts}dt.$$

Applying this transform to the differential equation leads to

$$\psi(z,s) = \frac{z^{n+1} - \mu(1-z)\pi(s)}{zs - (\mu - \lambda z)(1-z)}. \tag{3}$$

It is also convenient to discuss this equation in terms of the zeros of the denominator, regarded as a quadratic in z: These are denoted by ξ and η, where

$$\xi = \frac{\lambda + \mu + s - \left[(\lambda + \mu + s)^2 - 4\lambda\mu\right]^{1/2}}{2\lambda},$$

$$\tag{4}$$

$$\eta = \frac{\lambda + \mu + s + \left[(\lambda + \mu + s)^2 - 4\lambda\mu\right]^{1/2}}{2\lambda}.$$

In terms of these zeros, it is easy to show that

$$s = -\lambda(1-\xi)(1-\eta)$$

and

$$\xi\eta = \mu/\lambda = 1/\rho.$$

The first task is to evaluate $\pi(s)$, and so represent the transform $\psi(z,s)$ in terms of s, z, and the parameters λ, μ, and n. To do this, the student should note that the function is convergent for $z < 1$ and that ξ is the zero less than one. Therefore ξ must also be a zero of the numerator, and thus

$$\pi(s) = \frac{\xi^{n+1}}{\mu(1-\xi)}, \tag{5}$$

so that the explicit form for the transform of the probability generating function is

$$\psi(z,s) = \frac{z^{n+1} - (1-z)\xi^{n+1}/(1-\xi)}{-\lambda(z-\xi)(z-\eta)}.$$ (6)

It is not easy to invert this expression with respect to the variable s, and expansion in terms of z is also rather complicated. However, the student specializing in transforms and higher transcendental functions may be able to show from this equation that each of the probabilities can be expressed as a finite series of Bessel functions of imaginary argument.[†] Instead of pursuing the formal development further, some of the important special cases will be discussed in the next sections.

6.3. The $M_\lambda/M_\mu/1$ Queue: Oversaturation

When $\lambda > \mu$ (i.e., $\rho > 1$), the number present in the system tends to increase without limit, but does so in a stochastic manner, so that the possibility exists for decrease from time to time. The states of the system are all transient, in that the probability of a revisit tends to zero. Therefore the probability of the system being in any state tends to zero, and this fact can be inferred from the equations of Section 6.2.

It is not, however, useless to inquire about the evolution of the system. Depending on how great the discrepancy between λ and μ, the time spent in various states may vary considerably. For a given state x, define the indicator random variable (Section 1.5) $I(t)$:

$$I(t) = \begin{cases} 1 & \text{when } X(t) = x, \\ 0 & \text{otherwise.} \end{cases}$$

Then

$$\int_0^\infty I(t)\,dt$$

[†] Details can be found in COX, D. R., and MILLER, H. D. (1965), *The Theory of Stochastic Processes*, Methuen, London, p. 194 ff.

measures the total time spent in state x and

$$E\left(\int_0^\infty I(t)\,dt\right)$$

is the expected total time spent in state x. Call this quantity $T(x)$. Then

$$T(x)=\int_0^\infty E(I(t))\,dt. \tag{7}$$

But the expected value of $I(t)$ is only the probability $p_x(t)$ of the queue being in state x, so that

$$T(x)=\int_0^\infty p_x(t)\,dt.$$

Rather than evaluating this integral for each x, it is more natural to evaluate all of the integrals by finding

$$\int_0^\infty \phi(z,t)\,dt,$$

and this is the same as $\psi(z,0)$. Therefore an investigation of the expected duration in the transient states begins with setting $s=0$ in the formulas of Section 6.2.

With $s=0$, consider the equations for ξ and η. The radical becomes $|\lambda-\mu|$ and it is at this stage that the assumption $\rho>1$ comes into the picture, giving the values

$$\xi=1/\rho,$$

$$\eta=1.$$

Substituting these values into $\psi(z,0)$ gives

$$\psi(z,0)=\frac{\rho^{n+1}z^{n+1}-\rho^n z^{n+1}-1+z}{\lambda\rho^{n-1}(\rho z-1)(1-z)(\rho-1)}. \tag{8}$$

The factor $\rho z-1$ which appears in the denominator is also concealed in the numerator; removing it leaves

$$\psi(z,0)=\left(1+\sum_{j=1}^n (\rho-1)z^j\rho^{j-1}\right)\Big/\left[\lambda\rho^{n-1}(1-z)(\rho-1)\right]. \tag{9}$$

There is now no difficulty in expanding Eq. (9) in powers of z; the coefficients are precisely the $T(x)$:

$$T(x) = \frac{1}{\lambda \rho^{n-x-1}(\rho-1)}, \qquad x = 0, 1, \ldots, n,$$

$$T(x) = \frac{1}{\lambda-\mu}, \qquad x = n+1, n+2, n+3 \ldots. \tag{10}$$

Note that if the queue begins empty ($n=0$), the expected time spent in each state is $(\lambda-\mu)^{-1}$.

From a practical point of view, the main interest is in the case $\rho < 1$, since infinite queues are more important as a theoretical concept than as an observed reality. In the remainder of this chapter, except where particularly stated, it is assumed that $\rho < 1$.

6.4. The $M_\lambda/M_\mu/1$ Queue: Equilibrium

When $\rho < 1$, it has been shown already that the equilibrium distribution is geometric. It is not difficult to obtain this result from the formulation of Section 6.2. The values of ξ and η are respectively 1 and $1/\rho$. Therefore, using Theorem 2 of Section 4.11,

$$\lim_{t \to \infty} \phi(z, t) = \lim_{s \to 0} s\psi(z, s)$$

$$= \frac{-\mu \lim_{s \to 0} s\pi(s)}{-(\mu-\lambda s)}. \tag{11}$$

To evaluate $\lim s\pi(s)$, first note that, with L'Hôpital's rule,

$$\lim_{s \to 0} \frac{s}{1-\xi} = \mu-\lambda,$$

and thus $\lim s\pi(s) = 1-\rho$. Substituting this value gives

$$\lim_{t \to \infty} \phi(z, t) = \frac{1-\rho}{1-\rho z}. \tag{12}$$

A rather interesting fact about the $M/M/1$ queue in equilibrium is that the distribution of time between departures is independent of the rate of service. There are, in fact, two kinds of consecutive departures: those separated by a service period (when the second customer departing joined a nonzero queue) and those separated by a service period plus an interarrival period (when the second customer proceeded directly into service on arrival). These types are illustrated in Fig. 1.

In the first case, on the left in Fig. 1, the time between departures has density function $\mu e^{-\mu x}$, and the probability of this case is the probability that the queue is not empty. Thus the contribution to the Laplace transform is

$$\rho \frac{\mu}{\mu+s}.$$

In the second case there is also a service period, but it is preceded by an *idle period*, that is, a time when the server is idle and the queue is empty. By the memoryless property of the negative exponential, this period, ending in an arrival, has density $\lambda e^{-\lambda x}$. Thus the total length between departures has density

$$\lambda e^{-\lambda x} * \mu e^{-\mu x}, \tag{13}$$

and this case has probability $1-\rho$, so that the contribution to the Laplace transform is

$$(1-\rho)\frac{\lambda}{\lambda+s}\frac{\mu}{\mu+s}. \tag{14}$$

Adding together the two Laplace transform pieces gives the result $\lambda/(\lambda+s)$.

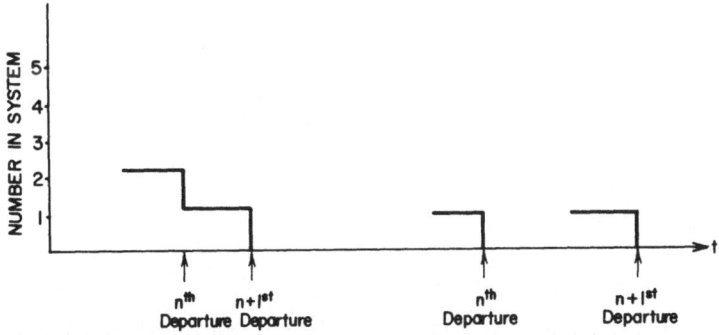

Figure 1. Two kinds of consecutive departures from a queue.

This fact is not so paradoxical as it is sometimes made to appear. It simply shows that a Poisson stream scrambled by another Poisson stream remains Poisson and retains the same parameter. The process is assumed to be in equilibrium, and so it is difficult to imagine what the output parameter would be, if not λ.

The case $\rho = 1$ may be more of a paradox. Since the states are all null recurrent, $p_x(t) = 0$ for all values of x. Yet, with a little modification of the argument in Section 6.3, it can be shown that (since $\xi = \eta = 1$) the mean time spent in each state is infinity.

6.5. The $M_\lambda / M_\mu / n$ Queue in Equilibrium: Loss Formula

There are many queueing variants on the $M_\lambda / M_\mu / 1$ queue as formulated in Sections 6.2–6.4. An item of information of great importance is the equilibrium distribution of number in the system, or, failing that, the mean and variance of the number in the system. The calculations needed to obtain the equilibrium distribution often follow rather closely the birth-and-death formulation and hardly require consideration of probability generating functions, Laplace transforms, and similar mathematical ornaments. In this section, an example is given of quick calculations leading to the equilibrium probabilities for a slight extension of the $M_\lambda / M_\mu / 1$ queue, namely, that where n servers are working. It is sensible to assume that the customers waiting are assigned to the first free server; otherwise the system is no more than n independent queues.

This means that the parameters in the birth-and-death scheme are λ for arrivals, and for departures,

$$\mu_x = \begin{cases} \mu x, & 0 \le x \le n, \\ \mu n, & x \ge n. \end{cases}$$

Thus the differential equations of the system are

$$p_0'(t) = -\lambda p_0(t) + \mu_1 p_1(t),$$

$$p_x'(t) = -(\lambda + \mu_x) p_x(t) + \lambda p_{x-1}(t) + \mu_{x+1} p_{x+1}(t). \tag{15}$$

Making the assumption of equilibrium, the left sides of these equations can be set equal to zero. Letting $p_x(\infty) = p_x$, that is, suppressing the time index in equilibrium, gives

$$\mu p_1 - \lambda p_0 = 0,$$

$$\mu(x+1)p_{x+1} - (\lambda + \mu x)p_x + \lambda p_{x-1} = 0, \qquad x = 1, 2, \ldots, n-1,$$

$$n\mu p_{x+1} - (\lambda + \mu n)p_x + \lambda p_{x-1} = 0, \qquad x = n, n+1, n+2, \ldots.$$

With some algebraic effort, these equations can be solved successively in terms of p_0 as follows:

$$p_x = \frac{\rho^x}{x!} p_0, \qquad x = 0, 1, \ldots, n,$$

$$p_x = \frac{\rho^x}{n^{x-n} n!} p_0, \qquad x = n+1, n+2, n+3 \ldots.$$

(16)

In the usual way, the value of p_0 can be obtained from the condition that the distribution normalizes to unity:

$$p_0 = \left(\sum_{j=0}^{n-1} \frac{\rho^j}{j!} + \frac{\rho^n}{n!(1-\rho/n)} \right)^{-1}, \qquad (17)$$

thus reducing the problem to one of calculation.[†]

An important special case arises when $n \to \infty$. If there are an infinite number of servers, there is no queue proper, since every customer in the system is being served. An example of such a service system would be an infinite parking lot, where the service time is equated to the parking time. It is left as an exercise for the student to show that in this case the probability distribution of the number in the system—the number of cars parked—is, in equilibrium, Poisson.

Finite Storage Space

In many cases it is more reasonable to assume that only a finite number of customers can be waiting in the queue. This would be true for a finite parking lot. It also applies to a problem which was of primary significance

---

[†]In this treatment, $\rho = \lambda/\mu$ as usual. However, some authors prefer to keep the symbol ρ to denote the traffic intensity of the queue, i.e., the ratio of input to service, so that the equilibrium condition remains $\rho < 1$. In the notation used here, the equilibrium condition is $\rho/n < 1$, since there are n servers at work.

in the development of queueing theory early in this century: the design of telephone switchboards. A telephone switchboard is built with a certain number of channels, just as a parking lot has a fixed number of slots. These are designed into the system at some cost, and not easily changed. Each call is placed in a channel, and when the channels are all occupied, further calls are "lost" to the system. Thus there are n "servers," a server being a channel or a slot, but no queue is permitted to form. In the design of the switchboard or parking lot, a level of service is prescribed as the probability that a call will be lost, or a car find the lot filled. Then, given the demand parameter λ and the occupancy parameter μ, the designer wishes to discover a relationship between the probability of a lost demand and the number of channels n so that n can be expressed in terms of λ, μ and the probability of a lost demand.

This is not a particularly difficult problem to solve. The student should fill in the details leading to the equilibrium probability equations

$$-\lambda p_0 + \mu p_1 = 0,$$

$$\lambda p_{x-1} - (\lambda + x\mu) p_x + (x+1)\mu p_{x+1} = 0, \qquad x = 1, 2, \ldots, n-1, \qquad (18)$$

$$\lambda p_{n-1} - n\mu p_n = 0,$$

which yield

$$p_x = \frac{1}{x!} \rho^x p_0, \qquad x = 0, 1, \ldots, n,$$

that is, the *truncated Poisson distribution*:

$$p_x = \frac{\rho^x / x!}{1 + \rho + \rho^2/2 + \cdots + \rho^n/n!}, \qquad x = 0, 1, \ldots, n. \qquad (19)$$

For the design of the switchboard or of the parking lot, the value p_n, the last value, is the important one, since it gives the probability that, in equilibrium, the facility will be full and a loss will occur. This formula,

$$p_n = P(\text{loss}) = \frac{\rho^n / n!}{1 + \rho + \rho^2/2 + \cdots + \rho^n/n!}, \qquad (20)$$

is known as *Erlang's*[†] *Loss Formula* and is rather famous. It has been tabulated, so that n can be obtained from λ, μ, and p_n. In practice, the first

[†]A. K. Erlang, Danish engineer, 1878–1929.

two of these parameters would be observed, and the third, p_n, set as a quality standard for the system.

The Poisson distribution for $M/M/\infty$ is also obtainable from this formulation by letting $n \to \infty$.

6.6. $M_\lambda/G_\mu/1$ Queue and the Imbedded Markov Chain

The method of differential difference equations, deriving from the backward equations of a birth-and-death process, can be used in queueing theory only when *both* elements in the first two places of the Kendall symbol are M. This is true because only in this case is it possible to analyze the possible transitions according to the Poisson process scheme of axioms of Section 5.3. As the preceding sections indicate, this can cover a considerable variety of queueing situations.

When one *or* the other of the first two places in the Kendall symbol is M, another useful technique can be used, the method of the imbedded Markov chain. In this section a general discussion is given of the $M_\lambda/G_\mu/1$. It is assumed that the arrival process is Poisson with parameter λ. The service time is denoted by U and has distribution function $B(x)$, density function $b(x) = B'(x)$, Laplace transform

$$\beta(s) = \int_0^\infty b(x)e^{-sx}dx,$$

and mean service interval

$$1/\mu = -\beta'(0),$$

with variance

$$v = \beta''(0) - [\beta'(0)]^2.$$

Service times are assumed to be independently equidistributed.

As soon as a customer enters service, the probability of service termination begins to change, in contrast with the M service case, which exhibits no aging. Whether the service distribution shows positive or negative aging, it is incorrect to assume that the aging is constant, except in the special case

$$\beta(s) = \frac{\mu}{\mu+s}.$$

It is therefore impossible to speak of the probability of a service termination in time Δt, since this probability depends not only on Δt, but also on how long the customer has been in service. One way around the difficulty is to examine the queue only when a service has just finished, that is, when aging is momentarily absent. Let the random variable X_n denote the number in the system immediately after the nth departure from the system. Consecutive values of X_n reflect one departure and a certain number of arrivals. But the arrivals are assumed to be Poisson, and so the probability of any given number of arrivals can be calculated.

Let q_x denote the probability of x arrivals during a service period. Then, conditional on a service period of length u, the distribution is by definition Poisson:

$$q_x = \frac{e^{-\lambda u}(\lambda u)^x}{x!} \qquad \text{for fixed } u.$$

Therefore, averaging over the distribution of U,

$$q_x = \int_0^\infty \frac{e^{-\lambda u}(\lambda u)^x}{x!} b(u)\, du. \tag{21}$$

The probability generating function[†] of this distribution can be written

$$\theta(s) = \sum_{j=0}^\infty q_j s^j$$

$$= \sum_{j=0}^\infty s^j \int_0^\infty \frac{e^{-\lambda u}(\lambda u)^j}{j!} b(u)\, du$$

$$= \int_0^\infty e^{-\lambda u + \lambda s u} b(u)\, du$$

$$= \beta[\lambda(1-s)]. \tag{22}$$

The probabilities q_x have already been discussed in some detail in Section 3.9 (although without reference to the imbedded aspect), the transition matrix for the imbedded Markov chain written down, the states

---

[†]See the footnote in Section 5.11.

classified, and the solution to the stationary case obtained:

$$
\begin{pmatrix}
q_0 & q_1 & q_2 & \cdots \\
q_0 & q_1 & q_2 & \cdots \\
0 & q_0 & q_1 & \cdots \\
0 & 0 & q_0 & \cdots \\
\vdots & \vdots & \vdots &
\end{pmatrix},
$$

leading to

$$
\phi(s) = \frac{\theta(s)(1-s)(1-\rho)}{\theta(s)-s}, \tag{23}
$$

where $\phi(s)$ is the probability generating function of the equilibrium distribution. The equilibrium distribution of the total number in an $M/G/1$ queue can be expressed directly in terms of the service distribution by a small substitution:

$$
\phi(s) = \frac{(1-s)(1-\rho)\beta[\lambda(1-s)]}{\beta[\lambda(1-s)]-s}. \tag{24}
$$

The student can verify that this formula is consistent with the $M/M/1$, result, that is, if

$$
\beta(s) = \frac{\mu}{\mu+s},
$$

then

$$
\phi(s) = \frac{1-\rho}{1-\rho s}.
$$

The $M/D/1$ Queue

When each service period has exactly the same length $(1/\mu)$,

$$
b(x) = \delta(x - 1/\mu),
$$

a deterministic distribution, with Laplace transform

$$
\beta(s) = e^{-s/\mu}.
$$

Therefore

$$\phi(s)=\frac{(1-s)(1-\rho)}{1-se^{\rho(1-s)}}$$

$$=(1-s)(1-\rho)\sum_{j=0}^{\infty}s^je^{j\rho(1-s)}. \tag{25}$$

To find values of p_x it is necessary to expand the exponential in powers of s and pick out the coefficient of s^x. Note first that the first two values are easy to find:

$$p_0=1-\rho,$$

$$p_1=(1-\rho)(e^\rho-1).$$

Therefore

$$\phi(s)-p_0-sp_1=-s(1-\rho)e^\rho+(1-s)(1-\rho)\sum_{j=1}^{\infty}\sum_{i=0}^{\infty}e^{j\rho}s^j\frac{(-s\rho j)^i}{i!},$$

and the coefficient of s^x is found to be (with the little algebraic attention to detail)

$$p_x=(1-\rho)\left(\sum_{j=0}^{x}\frac{(-\rho j)^{x-j}e^{\rho j}}{(x-j)!}-\sum_{j=0}^{x-1}\frac{(-\rho j)^{x-j-1}e^{\rho j}}{(x-j-1)!}\right),\qquad x=2,3,4,\ldots. \tag{26}$$

This is a fairly complicated formula for a rather simple queueing situation: Poisson arrivals, deterministic service, one server, equilibrium.

In the "bridge" case between deterministic and negative exponential service, $M/E_k/1$, it is easy enough to write down the generating function, but quite difficult to find a convenient expression for the probabilities.

Lacking good expressions for probabilities, it is natural that those wishing to apply the theory should resort to consideration of the mean and variance. This problem is considered in the following section.

6.7. The Pollaczek–Khintchine Formula

This well-known and useful formula expresses the mean number in the system in equilibrium in terms of the mean and variance of the service time distribution. In order to obtain the formula, it is only necessary to differentiate Eq. (23) of Section 6.6 and evaluate it at the value $s=1$, but the work requires two applications of L'Hôpital's rule, and so the calculation needs to be carefully organized. First, note that

$$\theta(1)=\beta(0)=1,$$

$$\theta'(1)=\rho,$$

$$\theta''(1)=\lambda^2\beta''(0)=\rho^2+\lambda^2v.$$

Then, differentiating,

$$\phi'(s)=\frac{(1-\rho)[-\theta^2-s\theta'+s^2\theta'+\theta]}{[\theta(s)-s]^2}.$$

For the substitution $s=1$, it is necessary to differentiate numerator and denominator separately twice to reduce the zero-over-zero indeterminacy. This yields the Pollaczek–Khintchine formula

$$E(X)=\phi'(1)=\frac{2\rho-\rho^2+\lambda^2v}{2(1-\rho)}, \tag{27}$$

which is often written in the form[†]

$$E(X)=\rho+\frac{\lambda^2v+\rho^2}{2(1-\rho)}. \tag{28}$$

It is easy to see from this formula that the minimum mean number in the system is achieved by setting the variance equal to zero, assuming a fixed service rate. This means that the $M/D/1$ queue yields less congestion than any other $M/G/1$ system.

An Alternative Proof

For the $M/G/1$ queue, the service times form a renewal process when service is taking place, and the arrivals occur at instants which have no relationship to service. Therefore the distribution of the time from an arrival

---

[†]A slight variant of this form relates to the mean waiting time rather than the mean number in the system; see Section 6.8.

to a service termination (assuming that someone is in service at the time of arrival) is given by Theorem 4 of Section 5.10, and the moments of this random variable, by the formula of Problem 22 of Section 5.15. In particular, the expected time from the arrival to the first service termination is obtained by setting $n=2$ in that formula and is

$$\text{expected residual service} = \frac{v + \mu^{-2}}{2\mu^{-1}} \rho,$$

where ρ is introduced as the probability that a customer is being served when the arrival takes place, i.e., that the queue is not empty. (It is important to realize that the formula $p_0 = 1 - \rho$ does not depend on the service distribution.) Let V be the time that the newly arrived customer must wait before proceeding into service, and let Y be the number of customers queueing, excluding the one being served. Then the expected value of V is the sum of one expected residual service and Y times $1/\mu$:

$$E(V) = \frac{\lambda^2 v + \rho^2}{2\lambda} + \frac{1}{\mu} E(Y).$$

It is intuitively clear that in equilibrium $E(Y) = \lambda E(V)$, since the expected number of arrivals in any period is the length of the period multiplied by the arrival rate.[†] Therefore

$$E(Y) = \frac{\lambda^2 v + \rho^2}{2(1 - \rho)}.$$

The relationship between $E(Y)$ and $E(X)$, the total number in the system, is not, as one might guess, $E(X) = E(Y) + 1$; this is true when the service is occupied. Otherwise $E(Y) = E(X)$. Thus

$$E(Y) = E(X)(1 - \rho) + \left[E(X) - 1\right]\rho$$

$$= E(X) - \rho, \tag{29}$$

which proves the Pollaczek–Khintchine formula.[‡]

[†] The validity of this formula is discussed in Section 6.8.

[‡] The four versions of the formula, applying to the random variables X, Y, V, and W, vary slightly; any one of them might be called the Pollaczek–Khintchine formula.

6.8. Waiting Time

This section begins the treatment of a second random variable associated with a queue, namely, the waiting time in the system. The *waiting time* of a customer is defined to be the time beginning with arrival into the system and ending with termination of service. The time is divided into two parts by the instant at which service begins: The first part is called the *queueing time* and the second, the *service time*. When the queue is empty on arrival, and the service mechanism is also empty, the arrival proceeds directly into service and therefore the queueing time is zero. Thus the queueing time distribution is mixed discrete and continuous, with a discrete component at the origin of magnitude equal to the probability that the number in the system is zero (for an $M/G/1$ queue, $1-\rho$). The distribution of waiting time consists of the convolution of this mixed distribution with the service time distribution.

Although the distributions of X, the number in the system, and W, the waiting time in the system, are quite different, they are connected by a simple formula[†] relating to the mean values:

$$E(X)=\lambda E(W).\tag{30}$$

The intuitive significance of the formula becomes clear by considering the following argument leading to it. During time t in statistical equilibrium, the expected number of arrivals to the system is λt. With expected waiting time $E(W)$, the expected total waiting time brought into the system in time t is $\lambda t E(W)$. If the expected number present in the system is $E(X)$, then the expected amount of waiting time consumed in the interval of length t is $E(X)t$. In equilibrium, the input of waiting time must equal its expenditure, and so

$$\lim_{t\to\infty}\lambda t E(W)=\lim_{t\to\infty}E(X)t,$$

from which the formula follows.

Although the relationship between $E(X)$ and $E(W)$ must appear quite obvious, and indeed nearly trivial, it has some interesting features: (i) It is

[†]This formula is usually known as $L=\lambda W$, where L and W are the respective expectations. There is an extensive literature on the subject: see, for example, *Operations Research* **9**, 383–387 (1961), **15**, 1109–1116 (1967); **17**, 915–917 (1969); **18**, 172–174 (1970); **20**, 1115–1126 (1972); **20**, 1127–1136 (1972); **22**, 417–421 (1974).

true for systems of very great generality, without specific assumptions as to the input and service process. The exact degree of generality is discussed in some references cited in the footnoted papers. (ii) It implies among other things that the mean wait must be independent of such considerations as the order of service, since the mean number in the system does not depend on them. (iii) It yields a version of the Pollaczek–Khintchine formula for expected waiting time simply by dividing Eq. (27) by λ. (iv) It applies equally to Y, the number queueing, and V, the time queueing:

$$E(Y)=\lambda E(V); \tag{31}$$

this provides an alternative proof of $E(Y)=E(X)-\rho$, since by definition $E(V)=E(W)-1/\mu$.

It is also possible to find a relatively simple relationship between the random variables

$$X(t)=\text{number in the system at time } t$$

and

$$W_n =\text{waiting time of the } n\text{th customer}$$

by using indicators. Let

$$I_n(t)=1 \qquad \text{if the } n\text{th customer is in the system at time } t,$$

$$I_n(t)=0 \qquad \text{otherwise.}$$

Then

$$X(t)= \sum_{j=1}^{\infty} I_j(t)$$

and

$$W_n = \int_0^{\infty} I_n(u)\,du.$$

Finding the distribution of W_n is a little more difficult and depends on the type of queue under consideration. In this and the next section, several examples will be given of arguments leading to the distribution of waiting time and queueing time. The queueing time V is a mixed discrete and continuous variable, with continuous portion $a(x)$, distribution function $A(x)$, and Laplace transform $\alpha(s)$.[†] The waiting time $W=U+V$, where U is the service time, has density $c(x)$, distribution function $C(x)$, and Laplace transform $\gamma(s)$. Since $a(x)$ represents only a portion of the probability for the queueing time, it normalizes not to unity, but to $P(X>0)$, which in the $M/G/1$ case is ρ.

Consider first the $M/M/1$ queue in equilibrium. Suppose an arriving customer is confronted with one customer being served and y customers queueing, where $y=0,1,2,\ldots$. Then the time this customer must queue before proceeding into service is the sum of y negative exponential variables. Thus, conditional on y, the distribution $a(x)$ is

$$(\mu e^{-\mu x})^{y*},$$

and since the number in the system is geometrically distributed with parameter ρ, the continuous portion of the queueing distribution can be written

$$a(x)= \sum_{j=1}^{\infty} (\mu e^{-\mu x})^{j*}(1-\rho)\rho^{j}$$

$$=\lambda(1-\rho)e^{-(\mu-\lambda)x}, \tag{32}$$

where the equilibrium assumption means that $\mu>\lambda$. The student can verify that this normalizes to ρ. Taking into consideration the discrete component at the origin of magnitude $1-\rho$, the Laplace transform can be written

$$\alpha(s)=\rho\frac{\mu-\lambda}{\mu-\lambda+s}+1-\rho$$

$$=\frac{(1-\rho)(\mu+s)}{\mu-\lambda+s}, \tag{33}$$

[†] The transforms are Laplace transforms of the densities and Laplace–Stieltjes transforms of the distribution functions.

so that the Laplace transform of the waiting time is

$$\gamma(s) = \beta(s)\alpha(s)$$

$$= \frac{\mu(1-\rho)}{\mu-\lambda+s}$$

$$= \frac{\mu-\lambda}{\mu-\lambda+s}. \tag{34}$$

Thus the waiting time distribution is simply negative exponential with parameter $\mu-\lambda$. It is easy to verify that the relationship $E(X)=\lambda E(W)$ holds in this case.

Next, consider an $M/G/1$ queue in equilibrium. Suppose a customer finishes service and leaves X other customers in the system after he departs. If his waiting time is W, then X is a Poisson random variable with parameter λW. Thus, letting $p_x = P(X=x)$, as usual, and $\phi(s)=\Sigma p_j s^j$,

$$p_x = \int_0^\infty e^{-\lambda u}(\lambda u)^x c(u)\, du,$$

so that

$$\phi(s) = \int_0^\infty e^{-\lambda u} c(u) \sum_{j=0}^\infty \frac{(s\lambda u)^x}{x!}\, du$$

$$= \int_0^\infty e^{-\lambda u} e^{s\lambda u} c(u)\, du$$

$$= \gamma[\lambda(1-s)]. \tag{35}$$

Breaking down waiting time into queueing and service times,

$$\phi(s) = \alpha[\lambda(1-s)]\beta[\lambda(1-s)], \tag{36}$$

which yields the following expression for the waiting time transform in terms of the service time transform and the probability generating function for the number in the system:

$$\alpha(s) = \frac{\phi[(\lambda-s)/\lambda]}{\beta(s)}. \tag{37}$$

Substituting for $\phi(s)$ the equilibrium value in terms of $\beta[\lambda(1-s)]$ obtained in Section 6.6 [Eq. (24)], the following result is obtained:

$$\alpha(s) = \frac{s(1-\rho)}{\lambda\beta(s)-\lambda+s}, \tag{38}$$

or

$$\gamma(s) = \frac{s\beta(s)(1-\rho)}{\lambda\beta(s)-\lambda+s}. \tag{39}$$

It is, of course, easy to verify that these formulas give the correct answer in the $M/M/1$ case.

A convenient form of $\alpha(s)$ can be obtained by considering the *residual service time* when a customer enters the queue, that is, the time from his entry, assuming there is service in process, until that service terminates. Let the residual service time be denoted by $\bar{U}$ with Laplace transform $\bar{\beta}(s)$. Then, noting Theorem 4, Section 5.10,

$$\alpha(s) = \frac{1-\rho}{1-\rho\bar{\beta}(s)}$$

$$= (1-\rho) \sum_{j=0}^{\infty} \left(\rho\bar{\beta}(s)\right)^j. \tag{40}$$

Thus the queueing time distribution is conveniently represented in terms of convolutions of the residual service time distribution. However, to obtain an explicit form for the queueing time distribution may involve a rather complicated Laplace inversion. For example, with a $M/D/1$ queue, it is necessary to perform Laplace inversion of powers of

$$\bar{\beta}(s) = \frac{\mu}{s}(1-e^{-s/\mu}) \tag{41}$$

(cf. Section 5.10, Theorem 5).

6.9. Virtual Queueing Time

The virtual queueing time $V(t)$ is defined to be the queueing time which would be experienced by a customer arriving at time t, whether or not

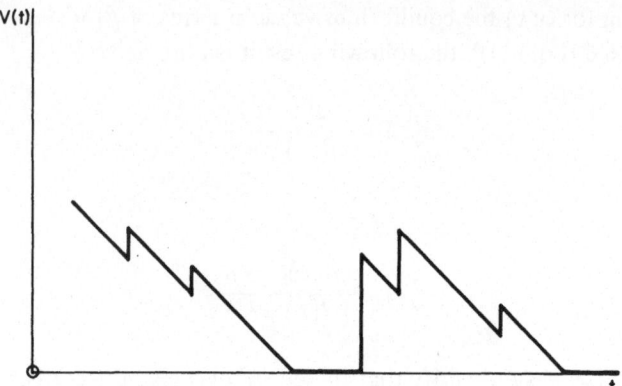

Figure 2. Virtual waiting time in a queue.

t is actually an arrival instant. If there is an arrival at time t, $V(t)$ is the queueing time of that customer, but if there is no arrival at time t, $V(t)$ is the total unexpired service required to clear the queue present at time t, namely, the sum of the service times of those queueing and the residual service time of the customer presently in service.

For an $M/G/1$ queue, the methods of Section 6.8 are not appropriate to the problem, since $V(t)$ must be considered at instants which are neither times of arrival or departure. Also, since aging in service is present (except for the $M/M/1$ case), the methods of Section 6.2 are unsuitable.

The solution to the problem of finding the distribution of queueing time lies in a consideration of the differential properties of its graph, which is shown in Fig. 2. Since the amount of work remaining in the system decreases with time (except when the server is idle), the graph declines at a 45° angle, except at arrival instants, when an increment of service is added. Using the notational system of Section 6.8, let[†]

$$A(x,t)=P(V(t)<x),$$

$$\alpha(s,t)=\int_0^\infty e^{-sx}\,dA(x,t).$$

[†] See the footnote in Section 5.11. Here the transform is with respect to x, not t.

Then, considering the graph of $V(t)$ at times t and $t+\Delta t$, it is easy to see that

$$A(x, t+\Delta t)=(1-\lambda\Delta t)A(x+\Delta t, t)+\lambda\Delta t$$

$$\times \int_0^{x+\Delta t} A(x-u+\Delta t, t)b(u)\, du+o(\Delta t).$$

The first term corresponds to the situation in which there is no arrival, and the second term to the situation where there is an arrival. Thus the first element M of the Kendall symbol is assumed. The equation is not as yet in the proper form for transformation into a differential equation, simply because the Δt is attached to one variable on the left side and a different variable on the right. This can be remedied, however, by use of the mean value theorem with respect to the variable x:

$$A(x+\Delta t, t)=A(x, t)+\frac{dA}{dx}\Delta t+o(\Delta t).$$

Substituting this value, it is now easy to form the following partial differential integral equation for A:

$$\frac{\partial}{\partial t}A(x, t)=\frac{\partial}{\partial x}A(x, t)-\lambda A(x, t)+\lambda\int_0^x A(x-u, t)b(u)\, du. \quad (42)$$

In taking the Laplace transform of this equation with respect to the variable x, it is necessary to remember [cf. Section 4.11, Eq. (37)] that in general $F(0+, t)\neq 0$. The transformation yields the equation

$$\frac{\partial}{\partial t}\alpha(s, t)=s\alpha(s, t)-sA(0+, t)-\lambda\alpha(s, t)+\alpha(s, t)\beta(s). \quad (43)$$

In equilibrium, the time derivative vanishes, leading to

$$\alpha(s)=\frac{sA(0+)}{\lambda\beta(s)-\lambda+s}. \quad (44)$$

The value $A(0+)=1-\rho$ is obtained by setting $s=0$ in the equation and evaluating the right side with the assistance of L'Hôpital's rule, bearing in mind that $\beta'(0)=-1/\mu$.

Thus in the $M/G/1$ case the probability distribution for the virtual queueing time is the same as that obtained for the actual queueing time. The reason is easy to see: For in an $M/G/1$ queue, arrivals, being "at random" in accordance with a Poisson process, occur at instants which are "typical" values of the virtual queueing time.

6.10. The Equation $y=xe^{-x}$

The various properties of the curve $y=xe^{-x}$ which are shown in Fig. 3 are easy to establish by elementary calculus: a single maximum at $(1,1/e)$, increasing from the origin to that point and decreasing from that point to infinity. It follows that for values of y greater than $1/e$, there are no real values of x, and for values of y less than $1/e$, there are two values of x: $x_1<1$ and $x_2>1$.

Theorem 1. If $y<1/e$ and $y=xe^{-x}$, where $x<1$, then

$$x= \sum_{j=1}^{\infty} \frac{j^{j-1}}{j!}y^j. \tag{45}$$

Proof. Let $x=\sum_{j=0}^{\infty}b_j y^j$. Clearly $b_0=0$, since $y=x-x^2+\frac{1}{2}x^3-\cdots$, and if there were a constant term in the expansion of x, there would also be one in the expansion of y: $b_0-b_0^2+\frac{1}{2}b_0^3-(1/3!)b_0^4+\cdots=b_0e^{-b_0}\neq0$.

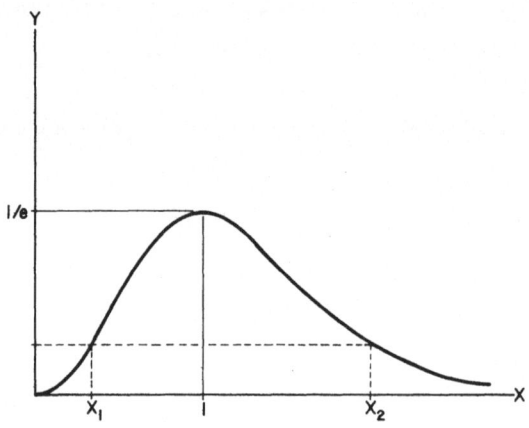

Figure 3. The curve $y=xe^{-x}$.

Therefore

$$x = \sum_{j=1}^{\infty} b_j y^j.$$

Differentiate with respect to x,

$$1 = \sum_{j=1}^{\infty} j b_j y^{j-1} \frac{dy}{dx}.$$

Let r denote a positive integer;

$$y^{-r} = \sum_{j=1}^{\infty} j b_j y^{j-r-1} \frac{dy}{dx}.$$

If $j \neq r$,

$$y^{j-r-1} \frac{dy}{dx} = \frac{1}{j-r} \frac{dy^{j-r}}{dx}$$

$$= \frac{1}{j-r} \frac{d}{dx} \left[x^{j-r} (A_0 + A_1 x + \cdots) \right],$$

and since x^{-1} is not the derivative of any power of x, there is no term in x^{-1}. If $j = r$, the term in the expansion is $r b_r y^{-1} dy/dx$, and since $dy/dx = y(1/x - 1)$, the term is $r b_r (1/x - 1)$. Thus the coefficient of $1/x$ in the expansion is $r b_r$. Therefore $j b_j$ is the coefficient of $1/x$ in the expansion of y^{-j} in powers of x. But

$$y^{-j} = (x e^{-x})^{-j}$$

$$= x^{-j} e^{xj}$$

$$= x^{-j} \sum_{i=0}^{\infty} \frac{(xj)^i}{i!}.$$

Thus

$$j b_j = j^{j-1} / (j-1)!,$$

which proves the theorem.   □

The student should examine the proof to see why it applies to the lesser value of x rather than to the greater value.

6.11. Busy Periods: Borel's Method

The discussion of queues has thus far dealt with two important random variables: the number in the system and the waiting time. A third variable of considerable practical importance is the duration of the busy period. This variable, together with the duration of the idle period, characterizes the operation of the system from the point of view of the server, rather than from the point of view of the customer.

Two preliminary remarks are necessary. First, the idle period in an $M/G/1$ queue is negative exponentially distributed with parameter λ, simply because it is terminated by an (Poisson) arrival instant. Second, the busy period can be defined to be either discrete or continuous. The *discrete busy period* is defined to be the number served between two consecutive idle periods: random variable K, probability h_x $(x=1,2,3,\dots)$, probability generating function $\kappa(s)$. The *continuous busy period* is defined as the time needed to serve these K customers: random variable L, density $f(x)$ $(0<x<\infty)$, Laplace transform $\sigma(s)$.

The $M/D/1$ Queue and Borel's[†] Method

Suppose a busy period begins with a single customer. Let n_1 be the number of arrivals during the service period of this customer; then, by the result given in Section 6.6, the probability generating function for n_1 is $\beta[\lambda(1-s)]$, which in the constant service time case reduces to a Poisson distribution with parameter ρ. Let the number of arrivals during the n_1 service periods be n_2. Since the sum of Poisson variables is Poisson, the distribution of n_2 is Poisson with parameter n_1. Continue this argument by letting n_3 be the number of arrivals during the n_2 service periods, so that its distribution is again Poisson with parameter n_2. Suppose the process terminates, that is, the busy period ends when for some k the n_k service periods contain no arrivals. Then the probability of a busy period containing exactly this configuration $n_1, n_2,\dots, n_k$ is

$$\frac{e^{-\rho}\rho^{n_1}}{n_1!}\frac{e^{-\rho n_1}(\rho n_1)^{n_2}}{n_2!}\cdots\frac{e^{-\rho n_{k-1}}(\rho n_{k-1})^{n_k}}{n_k!}e^{-\rho n_k}, \qquad (46)$$

[†] Emile Borel, French mathematician, 1871–1956. The work on the busy period distribution was published during the German occupation of France. Could wartime conditions have contributed a motive for studying queues?

since the last factor represents the probability of no Poisson arrivals during n_k service periods. Note that $K=1+n_1+n_2+\cdots+n_k$, so that the probability of this type of a busy period (that is, with n_1 following the first customer, etc.) can be written

$$\Omega \rho^{K-1} e^{-\rho K},$$

where Ω is a function of $n_1, n_2, \ldots, n_k$ only and does not depend on the value of ρ.

In order to find $h_x = P(K=x)$, it is necessary to sum all possible expressions of this form, subject to the condition that $K=1+\Sigma n_j$. Rather than bother with such complicated algebra, it will be equivalent to write

$$h_x = \Omega_x \rho^{x-1} e^{-\rho x}, \qquad x=1,2,3,\ldots \tag{47}$$

and determine the coefficients Ω_x so that h_x normalizes to unity (when $\rho < 1$). Borel did this by integration in the complex plane. It is also possible to use the result of Section 6.10 for the purpose. Let

$$\rho e^{-\rho} = z.$$

Then

$$\sum_{j=1}^{\infty} h_j = 1 = \frac{1}{\rho} \sum_{j=1}^{\infty} \Omega_j z^j,$$

or

$$\rho = \sum \Omega_j z^j.$$

But the theorem of Section 6.10 shows that, when $\rho < 1$,

$$\rho = \sum \frac{j^{j-1}}{j!} z^j.$$

Since a power-series expansion is unique, it follows that

$$h_x = \frac{x^{x-1}}{x!} \rho^{x-1} e^{-\rho x}, \qquad x=1,2,3,\ldots. \tag{48}$$

This expression is called *Borel's distribution*.

The argument can easily be adapted to consideration of the over-saturated case $\rho > 1$. Everything proceeds just as in the equilibrium case $\rho < 1$ until the assumption of normalization to unity. In this case, there is a nonzero probability that the busy period will never terminate, say h_∞, and so the normalization must be

$$\sum_{j=1}^{\infty} h_j = 1 - h_\infty.$$

Let ρ' be the value < 1 such that

$$\rho' e^{-\rho'} = \rho e^{-\rho},$$

so that ρ' has the series expansion given in Section 6.10, with $y = \rho e^{-\rho}$. Then

$$1 - h_\infty = \frac{1}{\rho} \sum \Omega_j z^j$$

$$= \frac{1}{\rho} \sum \Omega_j (\rho' e^{-\rho'})^j$$

$$= \frac{\rho'}{\rho} \sum \Omega_j (\rho')^{j-1} e^{\rho' j}$$

$$= \frac{\rho'}{\rho},$$

so that

$$h_\infty = 1 - \rho'/\rho. \tag{49}$$

From this formula it is clear that when $\rho = 1$, $h_\infty = 0$, and the Borel distribution holds.

The student should show that the mean of the Borel distribution is $(1-\rho)^{-1}$ and that the variance is $\rho(1-\rho)^{-3}$.

The $M/M/1$ Queue and Borel's Method

There is no difficulty in modifying the Borel argument to fit the $M/M/1$ case; indeed, there are fewer problems. The distribution of the numbers of arrivals in a service period turns out to be geometric with parameter $\rho/(1+\rho)$. The student should show this, either directly or by substitution into the $M/G/1$ result of Section 6.6. Thus the probability of, for example, n_2 arrivals during n_1 services will be of the negative binomial form

$$\binom{n_2+n_1-1}{n_2}\left(\frac{1}{1+\rho}\right)^{n_1}\left(\frac{\rho}{1+\rho}\right)^{n_2}, \tag{50}$$

leading to a distribution of the form

$$h_x = \frac{\rho^{x-1}}{(1+\rho)^{2x-1}}\Omega_x. \tag{51}$$

For the normalization argument, and those which follow, it is necessary to invert not the difficult function $y=xe^{-x}$, but the much simpler function

$$y = \frac{x}{(1+x)^2}.$$

This function enjoys the same necessary properties: increasing to a maximum at $x=1$ and then decreasing. Inverting the function merely involves solving a quadratic equation and everything proceeds nicely. The attentive student will note, however, that the Ω_x have already been found in Section 1.12 and the discrete busy period distribution is, for the equilibrium case, the Catalan distribution given in that section [Eq. (27)]. The analysis of the oversaturated case $\rho>1$ is also not difficult, using the properties of the analogous number ρ', defined by

$$\frac{\rho}{(1+\rho)^2} = \frac{\rho'}{(1+\rho')^2}.$$

The student should confirm that $\rho\rho'=1$ and hence that

$$h_\infty = 1-1/\rho. \tag{52}$$

A comparison of the Borel distribution (for $M/D/1$) with the Catalan distribution (for $M/M/1$) shows a rather interesting difference between the busy period random variable and the other two random variables which have been discussed. Since the Catalan and Borel distributions have the same mean value, there appears to be no advantage from the point of view of busy periods in regular service over negative-exponential service. However, the mean values of queue length and queueing time (and waiting time) are all smaller in the $M/D/1$ case than in the $M/M/1$ case.

What is the situation of these three random variables with respect to variance in the two cases?

6.12. The Busy Period Treated as a Branching Process: The $M/G/1$ Queue

It is probably clear that the numbers of customers denoted by $n_1, n_2, n_3 \ldots$ in Borel's proof are identical with the generations which were defined in Section 3.12 in connection with branching processes. In explicit terms, each customer arriving during the service time of another customer is considered as a "descendant" of the one being served. The basic quantity, the distribution of number of arrivals during a service period, which was denoted by b_x with probability generating function $\beta(s)$ in Chapter 3, is in this chapter denoted by q_x, with probability generating function $\theta(s)$. Thus, for an $M/M/1$ queue, $\theta(s)=[1+\rho(1-s)]^{-1}$, geometric with parameter $\rho/(1+\rho)$, and for an $M/D/1$ queue, $\theta(s)=e^{-\rho+\rho s}$, Poisson with parameter ρ. In both these cases the mean is ρ, so that the limit of generation size is zero.

In terms of branching chains, the queueing busy period problem is to find the probability distribution for the total number of descendants of a single progenitor. Referring to the discussion of nested generating functions given in Section 3.12 and Problem 37, Section 3.17, it is only necessary to substitute the random variable $1+X_1+X_2+\cdots+X_N$ for the random variable $X_1+X_2+\cdots+X_N$ to see that the total number of descendants (or customers in a busy period) up to and including the nth generation has a probability generating function $\kappa_n(s)$ satisfying

$$\kappa_n(s)=s\theta\big[\kappa_{n-1}(s)\big]. \tag{53}$$

The condition $\rho<1$ guarantees the existence of the limit

$$\lim_{n\to\infty}\kappa_n(s)=\kappa(s).$$

Using an argument completely analogous to that of Section 3.13 (for the probability of extinction), the required probability generating function for the number of customers in the discrete busy period must satisfy

$$\kappa(s)=s\theta\big[\kappa(s)\big]. \tag{54}$$

The $M/D/1$ Queue

$$\kappa(s)=se^{-\rho+\rho\kappa(s)},$$

leading to

$$\rho\kappa(s)e^{-\rho\kappa(s)}=\rho se^{-\rho},$$

which by Section 6.10 yields

$$\kappa(s)=\sum_{j=1}^{\infty}\frac{j^{j-1}}{j!}(\rho se^{-\rho})^{j}\frac{1}{\rho}, \tag{55}$$

which is the probability generating function of the Borel distribution.

The $M/M/1$ Queue

$$\kappa(s)=s\big[1+\rho-\rho\kappa(s)\big]^{-1},$$

leading to the quadratic

$$\kappa(s)-\rho\big[\kappa(s)\big]^{2}+\rho\kappa(s)-s=0,$$

with the solution

$$\kappa(s)=\frac{1}{2\rho}\Big\{1+\rho\mp\big[(1+\rho)^{2}-4\rho s\big]^{1/2}\Big\}, \tag{56}$$

which, as shown in Section 1.11, is the probability generating function of the Catalan distribution.

Returning to the $M/G/1$ case, it has been shown in Section 6.6 that a simple relationship exists between $\theta(s)$ and $\beta(s)$; therefore $\kappa(s)$ satisfies the following functional equation:

$$\kappa(s)=s\beta\big[\lambda-\lambda\kappa(s)\big]. \tag{57}$$

The mean discrete busy period for this queue can be shown to agree with that for the $M/M/1$ and $M/D/1$ queues, either by differentiation or by the following direct argument.

Consider an entire cycle from the instant when the system becomes empty to the next instant when it becomes empty. This cycle consists of one busy period and one idle period for the server. In this context "busy period" means, of course, "continuous busy period." The time spent in equilibrium in the two states, busy and idle, will be proportional to the probabilities that the server is busy or idle, and these, from Section 6.6, are respectively ρ and $1-\rho$. Furthermore, the expected length of the idle period must be $1/\lambda$, since it is the mean of a negative exponentially distributed random variable with parameter λ. Thus the mean length of the continuous busy period Ω satisfies

$$\frac{\Omega}{1/\lambda} = \frac{\rho}{1-\rho},$$

so that

$$\Omega = \frac{1}{\mu} \frac{1}{1-\rho},$$

or, for the discrete busy period, the mean will be

$$\frac{1}{1-\rho}.$$

6.13. The Continuous Busy Period and the $M/G/1$ Queue

Although closely connected with the discrete busy period, the continuous busy period L has a less convenient distribution, even in the simplest cases. For example, in the $M/M/1$ case, the distribution $f(x)$ of L involves Bessel functions. It is not difficult, however, to derive a functional equation in the Laplace transform $\sigma(s)$ of the continuous busy period distribution, which is similar to Eq. (44) of Section 4.13.

The argument begins with conditioning on a fixed value u for the service time of the first arriving customer, and then on a fixed value j for the number of customers arriving during u. The distribution of u is $b(x)$, and that of j is Poisson with parameter λu. With these quantities fixed, the length of the busy period equals the combined lengths of each of the j busy periods induced by the j arrivals. This means that the busy period distribution, conditional on j and u, is the j-fold convolution of $f(x)$ with itself.

However, since x must be at least as large as u, the argument must be $x-u$. Thus the distribution of L can be written

$$f^{j*}(x-u), \qquad \text{conditional on } j \text{ and } u,$$

or

$$\sum_{j=0}^{\infty} \frac{(\lambda u)^j}{j!} e^{-\lambda u} f^{j*}(x-u), \qquad \text{conditional on } u,$$

or, unconditionally,

$$f(x) = \int_0^{\infty} b(u) \sum_{j=0}^{\infty} \frac{(\lambda u)^j}{j!} e^{-\lambda u} f^{j*}(x-u) \, du. \tag{58}$$

This equation, like so many other similar ones, can best be unraveled with the assistance of a Laplace transform. Since the transform of f^{j*} is σ^j, the transform of $f^{j*}(x-u)$ is $e^{-us}\sigma^j(s)$. Therefore

$$\sigma(s) = \sum_{j=0}^{\infty} \int_0^{\infty} \frac{(\lambda u)^j}{j!} e^{-\lambda u} e^{-us} \sigma^j(s) b(u) \, du$$

$$= \int_0^{\infty} \exp\left[-us - u\lambda + u\lambda\sigma(s) \right] b(u) \, du$$

$$= \beta\left[s + \lambda - \lambda\sigma(s) \right], \tag{59}$$

a fairly simple functional equation. Nevertheless, the direct solution must be deferred to the following section,[†] where some more general theory will be developed. Meanwhile, the student should verify the special cases $M/M/1$ and $M/D/1$ as follows.

The techniques of solution in these cases are exactly the same as already encountered in dealing with the discrete busy period: namely, the quadratic formula and the Lagrange series, Eq. (45). The results are fairly easy to confirm: For the $M/M/1$ case,

$$\sigma(s) = (1/2\lambda)\left\{ \lambda + \mu + s - \left[(\lambda + \mu + s)^2 - 4\lambda\mu \right] \right\}, \tag{60}$$

[†] However, students familiar with the Lagrange series [see WHITTAKER, E. T. and WATSON, G. N. (1927), *A Course of Modern Analysis*, 4th ed., Cambridge University Press, pp. 132–133] may be able to obtain the solution directly.

the transform of which yields a Bessel function for $f(x)$, and in the $M/D/1$ case,

$$\sigma(s) = \frac{1}{\rho} \sum_{j=1}^{\infty} \frac{j^{j-1}}{j!} \left(\rho e^{-\rho} e^{-s/\mu} \right)^j, \tag{61}$$

which transforms into

$$f(x) = \frac{1}{\rho} \sum_{j=1}^{\infty} \frac{j^{j-1}}{j!} \rho^j e^{-\rho j} \delta(x - j/\mu). \tag{62}$$

Thus, as would be expected in an $M/D/1$ queue, the distribution of the continuous busy period is hardly "continuous," being concentrated at the lattice points which are multiples of the mean service time $1/\mu$.

6.14. Generalized Busy Periods

This section is concerned with two ways to generalize the *basic* busy period distributions h_x and $f(x)$, a *discrete* generalization and a *continuous* generalization. Each of the ways of generalizing has some interesting applications, and each may be applied either to h_x or to $f(x)$. Table 6.1 on p. 228 will help to keep the terminology and notation clear.

Discrete generalization with parameter r, $r = 1, 2, 3, \ldots$. It is assumed that the busy period begins not with a single customer, but with an accumulation of r customers in the system. Thus the basic busy periods are obtained by setting $r = 1$.

Continuous generalization with parameter τ, $0 < \tau < \infty$. It is assumed that there is an initial period of duration τ during which customers accumulate and wait but are not served. Then the busy period begins with the number of customers who have arrived during τ as the value of r for the discrete generalization. It is not correct, however, to assume that the basic busy period distributions would be obtained by setting $\tau = 0$ (or any other value), since the basic busy periods are started not after any particular length of time, but rather when the first arrival occurs.

The discussion deals first with discrete busy period distributions (both generalizations) and then with continuous busy period distributions (both generalizations).

In both cases, however, it is the discrete generalization which provides the key to the continuous generalization, because the continuous generalization for a Poisson input queue requires only that the discrete generalization be weighted with the number of customers arriving during time τ, and this will be Poisson distributed.

The Discrete Busy Period

For the discrete generalization, beginning with r in the system, let h_x^{r*} denote[†] the probability that the busy period will consist of x customers, $x = r, r+1, r+2, \ldots$, and let $\pi_x(\lambda t)$ denote the Poisson probabilities with parameter λt, that is, the probability of x arrivals to the system in time t.

Theorem 1. For a $M/G/1$ queue,

$$h_x^{r*} = \frac{r}{x} \int_0^\infty \pi_{x-r}(\lambda t) b^{x*}(t)\, dt. \tag{63}$$

Proof (Tanner's[‡] Combinatorial Proof). In order for there to be x customers in a busy period, exactly $x-r$ arrivals must take place during the x service periods constituting the busy period, and these arrivals must be so placed in the service times that the queue does not become empty before the end of the x service periods. The first probability (of exactly $x-r$ arrivals) is given by the integral in the formula, and so it remains to show that the probability that these *arrivals occur in such a way that the queue never becomes empty* before the elapse of x service periods is r/x. This means that the total number in the system at the end of the xth service time must be less than it has ever been before, and this simple observation is the key to the argument.

Consider a particular pattern of arrivals, say a_1 arrivals in the first service time, a_2 arrivals in the second, and so forth. This particular pattern will be considered together with $x-1$ other patterns formed by permuting the pattern cyclically, that is,

$$a_1, a_2, \ldots, a_x$$

[†] Why is this distribution actually the r-fold convolution of h_x?
[‡] John Tanner, English mathematician, 1927–   .

together with

$$a_2, a_3, \ldots, a_1,$$

$$a_3, a_4, \ldots, a_2,$$

$$\vdots$$

$$a_x, a_1, \ldots, a_{x-1}.$$

The proof of the theorem consists in now showing that exactly r of each group of x patterns is "admissible" (i.e., leads to a busy period of x customers), assuming that the total number of arrivals is $x - r$.

The situation for $x = 10$, $r = 3$ is shown in Fig. 4, and the proof is keyed to that figure. In order to obtain all the cycles, the original pattern is assumed to be repeated once; then beginning with each of the first x service periods is equivalent to considering each of the x cycles. It is convenient to start with a queue of n (a large number) to illustrate all the cycles, since all

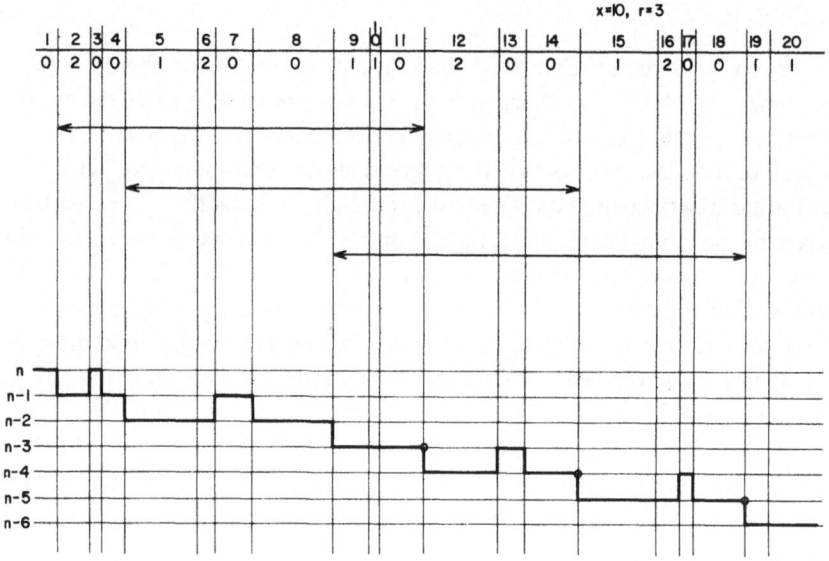

Figure 4. Pattern of arrivals and departures from a queue. The upper row of numbers represents the service times, the lower row, the number of arrivals during a service time. Arrows indicate admissible patterns. Adapted from TANNER, J. C. (1961), "A derivation of the Borel distribution," *Biometrika* **48**, pp. 222–224.

that need be shown is whether or not a cycle ends with a smaller number in the system than ever before.

It is clear that at the beginning of the $(x+1)$st service, the number in the system must be $n-r$, since there have been $x-r$ arrivals and x completed services. Similarly, at the beginning of each service period in the second half of the pattern, the number in the system is just r less than it was x services earlier.

Therefore, since the number of steps down is exactly r during the first half of the pattern, it is exactly r during the second half, and must on exactly r occasions take on a value lower than any previous value at the beginning of a service period. □

This concludes the proof of the theorem by Tanner's combinatorial method. In the special cases $M/M/1$ and $M/D/1$, proofs similar to Borel's proof (Section 6.11) can be constructed, except for the evaluation of the coefficients, which must then be done, as Borel did, by integration in the complex plane, rather than by use of Section 6.10. These special results are given in the following theorems; the proofs, specializing Theorem 1, are left as exercises for the student.

Theorem 2 (Borel–Tanner Distribution). For an $M/D/1$ queue,

$$h_x^{r*} = \frac{r}{(x-r)!} x^{x-r-1} e^{-\rho x} \rho^{x-r}, \qquad x=r, r+1, r+2,\ldots. \tag{64}$$

Theorem 3. For an $M/M/1$ queue,

$$h_x^{r*} = \frac{r}{x}\left(\frac{2x-r-1}{x-1}\right)\frac{\rho^{x-r}}{(1+\rho)^{2x-r}}, \qquad x=r, r+1, r+2,\ldots. \tag{65}$$

Turning now to the continuous generalization, suppose a time τ has elapsed without service starting and then the service mechanism begins to operate and produces a discrete busy period of x. First, in contrast to the discrete generalization, there is a nonzero probability of a zero busy period, corresponding to the probability that there are no arrivals during τ, namely, $e^{-\lambda\tau}$. The following theorems are not difficult to prove and are left as exercises for the student.

Let the distribution of the continuous generalization be $h_x(\tau)$. Then the relationship between $h_x(\tau)$ and h_x^{r*} is given as follows.

Theorem 4. For an $M/G/1$ queue,

$$h_x(\tau)= \sum_{j=1}^{x} \frac{e^{-\lambda\tau}(\lambda\tau)^j}{j!}h_x^{j*}, \qquad x=1,2,3,\ldots,$$

$$h_0(\tau)=e^{-\lambda\tau}. \tag{66}$$

Theorem 5. For an $M/G/1$ queue,

$$h_x(\tau)= \frac{e^{-\lambda\tau}\lambda^x\tau}{x!} \int_0^\infty e^{-\lambda u}b^{x*}(u)(u+\tau)^{x-1}\,du, \qquad x=1,2,3,\ldots,$$

$$h_0(\tau)=e^{-\lambda\tau}. \tag{67}$$

The proof consists simply in applying Theorem 1 to Theorem 4.

Theorem 6. For an $M/D/1$ queue,

$$h_x(\tau)=\frac{1}{x!}\lambda\tau e^{-\lambda\tau-\rho x}(\rho x+\lambda\tau)^{x-1}, \qquad x=1,2,3,\ldots,$$

$$h_0(\tau)=e^{-\lambda\tau}. \tag{68}$$

Theorem 7. For an $M/M/1$ queue,

$$h_x(\tau)= \frac{e^{-\lambda\tau}\tau\lambda^x\mu^x}{x!(x-1)!}\mathcal{L}_{\lambda+\mu}[u(u+\tau)]^{x-1}, \qquad x=1,2,3,\ldots,$$

$$h_0(\tau)=e^{-\lambda\tau}. \tag{69}$$

Note. Evaluation of this transformation leads to a rather complicated function, a modified Bessel function of the third kind.

Continuous Busy Periods

For the discrete generalization based on an accumulation of r customers before service begins, the continuous busy period distribution, like the discrete busy period distribution, is obviously a convolution of the basic busy period distribution and can therefore be written $f^{r*}(x)$. It is not difficult to express $f^{r*}(x)$ in terms of $b^{j*}(x)$ using Tanner's combinatorial method.

Theorem 8. For an $M/G/1$ queue,

$$f^{r*}(x) = \sum_{j=r}^{\infty} \frac{r}{j} \frac{(\lambda x)^{j-r}}{(j-r)!} e^{-\lambda x} b^{j*}(x). \tag{70}$$

Proof. The first term of this series represents the probability that there are no arrivals during the time required to serve the r accumulated customers, multiplied by the length of the r service periods. It is thus the value of $f^{r*}(x)$ conditional on no arrivals in the first r service periods. The same remark, reading j arrivals in place of no arrivals, holds for the jth term, with the additional factor r/j being the probability that the arrival pattern is admissible in the sense of Tanner's combinatorial argument.   ☐

Corollary 1. The solution to the functional equation (59) of Section 6.13 is

$$f(x) = \sum_{j=1}^{\infty} \frac{1}{j!} (\lambda x)^{j-1} e^{-\lambda x} b^{j*}(x). \tag{71}$$

Now, let the continuous generalization of the continuous busy period, that is, the time to finish a busy period started by an accumulation of customers during time τ, have a distribution denoted by $f(x, \tau)$.

Theorem 9. For an $M/G/1$ queue, $f(x, \tau)$ is a mixed discrete and continuous distribution with

$$f(x, \tau) = \sum_{n=1}^{\infty} b^{n*}(x) \frac{\lambda^n \tau}{n!} e^{-\lambda \tau - \lambda x} (x + \tau)^{n-1}, \qquad 0 < x < \infty,$$

$$\tag{72}$$

$$f(0, \tau) = e^{-\lambda \tau}.$$

Proof. The discrete portion of this distribution is obvious and clearly corresponds to $h_0(\tau)$. The continuous portion is obtained from the formula

$$f(x, \tau) = \sum_{n=1}^{\infty} \pi_n(\lambda \tau) f^{n*}(x).$$

This formula is quite analogous to the statement of Theorem 4, except for the fact that the upper limit of summation is infinity, rather than x. The reason for the change is that, beginning with j in queue, no less than j can be served in a busy period, but beginning with an elapse of time τ, there is no lower limit on the number which can be served in the busy period and, therefore, no upper limit on n. Substituting the value of $f^{n*}(x)$ given by Theorem 8 into this formula yields a double sum involving various convolutions of $b(x)$. Collecting the terms involving the nth convolution, it can be seen without too much trouble that the coefficient of $b^{n*}(x)$ is

$$\sum_{i=1}^{n} \frac{i}{n}\pi_i(\lambda\tau)\pi_{n-i}(\lambda x),$$

which can be simplified to yield

$$\frac{\lambda^n\tau}{n!}e^{-\lambda\tau-\lambda x}(x+\tau)^{n-1},$$

which is equivalent to the statement of the theorem.   □

There is not much profit in investigating the special cases $M/M/1$ and $M/D/1$ into Theorems 8 and 9, since the results do not simplify to any significant degree. For the $M/M/1$ case, however, Theorem 8, with $r=1$, does specify as an infinite series the Bessel function mentioned in Section 6.13. Also, the student would find it instructive to show that the expression for $f(x)$ in the $M/D/1$ case does agree with the similar result from Theorem 8. Hint: For any function $\Omega(x)$, the functions $\Omega(x)\delta(x-p)$ and $\Omega(p)$ $\delta(x-p)$ are identical.

Theorem 10. For an $M/G/1$ queue, the Laplace transform $\sigma(s, \tau)$ of $f(x, \tau)$ can be expressed in terms of $\sigma(s)$ by the formula

$$\sigma(s,\tau)=e^{-\lambda\tau+\lambda\tau\sigma(s)}. \tag{73}$$

Proof

$$f(x,\tau)=\sum_{j=1}^{\infty}\frac{e^{-\lambda\tau}(\lambda\tau)^j}{j!}f^{j*}(x)+\delta(x)e^{-\lambda\tau},$$

$$\sigma(s,\tau)=\sum_{j=0}^{\infty}\frac{e^{-\lambda\tau}(\lambda\tau)^j}{j!}\sigma^j(s),$$

which gives the result.   □

Note. In the terminology of Section 4.7, the $f(x, \tau)$ distribution is obtained by Poisson-mixing (parameter $\lambda\tau$) the $f(x)$ distribution.

6.15. The $G/M/1$ Queue

Up to this point the emphasis has been on queues with Poisson arrival processes. The reasons for this approach are two: (i) In the largest categories of applications, the assumption provides a good model for reality. Arrivals to a queue are often uncontrolled—they appear from no organized pool and can be well approximated by a Poisson stream. (ii) By treating queues of this type, it has been possible to illustrate a number of the basic techniques: differential difference equations, imbedded Markov chains, branching processes, transforms and generating functions, and so forth. By the judicious application of these techniques, it is possible to solve many other types of queueing configurations suggested by practical problems.

In some respects, the queue with Kendall symbol $G/M/1^{\dagger}$ is an image of $M/G/1$, but it has one striking difference: The equilibrium probabilities for the imbedded Markov chain of the number of customers in the system are geometric, independent of the input distribution. The interarrival assumptions, in fact, determine only the value of the geometric parameter. In order to show this, it is necessary to go through the usual steps of defining and solving the imbedded chain.

In a $G/M/1$ queue, the aging takes place between one arrival and another, and so the regeneration points are instants when a customer arrives. Suppose the interarrival density function is $d(x)$ with Laplace transform $\delta(s)$ and mean $-\delta(0)=1/\lambda$. Analogously to the q_x of Section 6.6, let r_x denote the probability of x service terminations during one interarrival gap, *assuming that the server is never free during this period.* This expression in italics is unnecessary in the $M/G/1$ definition of q_x by definition, and the lack of exact symmetry between the cases first appears here. Let the probability generating function of the r_x be $\eta(s)$. Then it is easy to express $\eta(s)$ in terms of $\delta(s)$:

$$r_x = \int_0^\infty \frac{(\mu t)^x}{x!} e^{-\mu t} d(t)\, dt,$$

so that

$$\delta(s) = \int_0^\infty e^{-\mu t + s\mu t} d(t)\, dt = \delta(\mu(1-s)). \tag{74}$$

---

†Some authors use GI in place of G; this renewal process is also called "Palm input."

The student should check carefully to see where the italicized assumption is required in this calculation.

In writing the transition matrix for the imbedded chain, every element can be expressed as one of the r_x, except for those in the first column. The reason for this is simply that a transition from any value x to zero represents a period (between one arrival and the next) which contains a certain time when the queue is empty, and so violates the italicized condition. Nevertheless, these probabilities can be filled in simply by assuming normalization of rows to unity. The student, taking into consideration these principles, should show that the transition matrix can be written

$$\begin{pmatrix} \sum_{1}^{\infty} r_j & r_0 & 0 & 0 & 0 & \cdots \\ \sum_{2}^{\infty} r_j & r_1 & r_0 & 0 & 0 & \cdots \\ \sum_{3}^{\infty} r_j & r_2 & r_1 & r_0 & 0 & \cdots \\ \vdots & \vdots & \vdots & \vdots & \vdots & \end{pmatrix}.$$

Before writing down and solving for the equilibrium probabilities defined by this matrix, it is first necessary to take a small excursion. Consider r_x as a probability distribution. The mean value is easily calculated:

$$\eta'(s)=\delta'(\mu(1-s))(-\mu),$$

$$\eta'(1)=\delta'(0)(-\mu)=(-1/\lambda)(-\mu)=1/\rho.$$

Thus, for the equilibrium condition $\rho<1$, the mean is >1, and the result of Section 1.13 shows that the equation

$$s=\eta(s)$$

has a unique root <1. Let this root be denoted by ζ.

Theorem 1. With the definitions given above, the matrix has stationary vector $(1-\zeta)\zeta^x$, $x=0,1,2,\dots.$

Proof.[†]   It is carried out by substitution into the equations defining the stationary vector. The first equation would be

$$p_0 = (1-\zeta)\left(\sum_1^\infty r_j + \zeta\sum_2^\infty r_j + \zeta^2\sum_3^\infty r_j + \cdots\right)$$

$$= (1-\zeta)\left[r_1 + r_2(1+\zeta) + r_3(1+\zeta+\zeta^2) + \cdots\right]$$

$$= (1-\zeta)\left[r_1 + (1-\zeta^2)r_2 + (1-\zeta^3)r_3 + \cdots\right]$$

$$= 1 - \eta(\zeta)$$

$$= 1 - \zeta.$$

Similarly, for the xth equation,

$$p_x = \sum_{j=x+1}^\infty p_j r_{j-x+1}$$

$$= (1-\zeta)\sum_{j=x+1}^\infty \zeta^j r_{j-x+1}$$

$$= (1-\zeta)\sum_{j=0}^\infty \zeta^{j+x-1} r_j$$

$$= (1-\zeta)\zeta^{x-1}\sum_{j=0}^\infty \zeta^j r_j$$

$$= (1-\zeta)\zeta^x. \qquad \square$$

Theorem 2. The asymptotic waiting time distribution is the following mixed discrete and continuous distribution:

$$(1-\zeta)\delta(x) + (1-\zeta)\zeta\mu e^{-\mu x(1-\zeta)}.$$

Proof. The proof is left as an exercise. [Note the difference between the impulse function $\delta(x)$ in this theorem and the similarly designated Laplace transform $\delta(s)$.] $\square$

[†] With considerably more difficulty, it is possible to prove this by "discovering" the geometric probabilities, rather than by verifying them.

This completes the treatment of queues which is based on combinations of the Kendall symbol. Although many of the basic techniques have been covered, it must be obvious that there are many types of queues which have been neglected. In the remainder of this chapter, three important queueing variants are treated: one altering arrival pattern, one changing service policy, and the third reversing queueing order. These give a small sample of how the basic techniques are used to solve the wide variety of queueing problems suggested by physical situations.

6.16. Balking

The name balking is applied to the input to a queue in which a customer on arrival makes a decision whether or not to join the queue, depending on his evaluation of whether the queue is too long or not. It is assumed that, observing the total number in the system $X(t)$, the customer compares this value with another integer X', which is the longest queue he is willing to tolerate. If $X(t) > X'$, he *balks*, that is, he goes away and does not return, while if $X(t) \leq X'$, he joins the queue and waits for service. It is furthermore assumed that the values of X' are distributed in the client population by means of some probability distribution

$$g_x = P(X'=x), \qquad G_x = P(X'<x), \qquad H_x = P(X'\geq x).$$

This is called the *balking distribution*.

With this model it is not difficult to build up the differential difference equations for an $M/M/1$ queue. Consider terms corresponding to: (i) no arrivals and no departures in Δt, (ii) one departure in Δt, (iii) one arrival who joins, and (iv) one arrival who balks. Let the parameter λ refer to the rate of arriving at the queue, including both those who join and those who balk. Then, with the usual argument,

$$p_x(t+\Delta t) = \left[1 - (\lambda+\mu)\Delta t\right] p_x(t) + \mu p_{x+1}(t)\Delta t$$

$$+ \lambda p_{x-1}(t) H_{x-1} \Delta t + \lambda p_x(t) G_x \Delta t + o(\Delta t), \qquad x = 1, 2, 3, \ldots, \qquad (75)$$

$$p_0(t+\Delta t) = (1 - \lambda\Delta t)p_0(t) + \mu p_1(t)\Delta t + o(\Delta t).$$

It is naturally assumed that no customer will balk at a completely empty system, i.e., $H_0 = 1$.

This leads to the system

$$p_0'(t) = -\lambda p_0(t) + \mu p_1(t),$$

$$p_x'(t) = -(\lambda + \mu)p_x(t) + \mu p_{x+1}(t) + \lambda p_{x-1}(t)H_{x-1} + \lambda p_x(t)G_x,$$

$$x = 1, 2, 3, \ldots,$$

which, in equilibrium, gives the recurrence relation

$$p_x = \rho p_{x-1} H_{x-1}, \qquad x = 1, 2, 3, \ldots,$$

with solution

$$p_x = \rho^x c_x p_0, \qquad p_0 = \left(\sum_{j=0}^{\infty} c_j \rho^j \right)^{-1},$$

where

$$c_x = \prod_{j=1}^{x-1} H_j, \qquad x = 2, 3, 4, \ldots, \quad c_0 = c_1 = 1.$$

Example 1 (Geometric Balking). Let

$$g_x = (1 - r)r^x.$$

Then

$$H_x = r^x$$

and

$$c_x = r^{x(x-1)/2},$$

so that

$$p_x = \rho^x r^{x(x-1)/2} p_0,$$

where[†]

$$p_0 = \left(\sum_{j=2}^{\infty} \rho^j r^{j(j-1)/2} \right)^{-1}.$$

Example 2 (Deterministic Balking). In some systems each customer has exactly the same value of X', say, the constant M. Then $g_M = 1$ and

$$H_x = 1, \qquad x = 0, 1, \ldots, M,$$

$$H_x = 0, \qquad x = M+1, M+2, M+3, \ldots.$$

Therefore

$$c_x = 1, \qquad x = 0, 1, \ldots, M+1,$$

$$c_x = 0, \qquad x = M+2, M+3, M+4, \ldots,$$

so that

$$p_0 = \frac{1-\rho}{1-\rho^{M+2}}$$

and

$$p_x = \rho^x \frac{1-\rho}{1-\rho^{M+2}}, \qquad x = 0, 1, \ldots, M+1.$$

It is easy to see in this example that the geometric distribution of X for an $M/M/1$ queue (Section 6.4) results when $M \to \infty$. With respect to parameters, it is important to note that since λ is not the joining rate, $\rho = \lambda/\mu$ is not the true ratio of arrivals to departures. This true rate is in fact the product of ρ with the probability of joining, that is,

$$\rho \left[1 - (p_1 G_1 + p_2 G_2 + \cdots) \right] = 1 - p_0.$$

It is left as an exercise for the student to discuss generating functions, means, and variances. In particular, show that for deterministic balking, the

---

[†] This kind of function is known as a false theta function.

mean number in the system in equilibrium is

$$\frac{\rho(1-\rho^{M+1})-(1-\rho)(M+1)\rho^{M+2}}{(1-\rho)(1-\rho^{M+2})}. \tag{76}$$

This too yields the classical result when $M \to \infty$.

Balking is an example of a structural variant in queueing theory which relates to the entry process. It is not too difficult to think of various other assumptions which might be appropriate in specific circumstances, for example, periods when entry is forbidden. However, in general, the arrival process is relatively out of the control of the engineer, while the service process is more subject to application of policy. It is therefore in variants of service that useful applications can be most often found. Some of these possibilities are discussed in the following section.

6.17. Priority Service

A number of interesting and useful variations in the service mechanism of a queue are comprised under the general heading of *priority* service. The priority classes may be intrinsic to the customers in some way, or they may be assigned by a manager of the queueing system in an effort to reduce, for example, overall delay.

Priority disciplines are classified first as being either *preemptive* or *head-of-the-line*. Preemptive service means simply that an arrival of higher[†] priority than the customer being served proceeds immediately into service, while head-of-the-line means that the customer in service is permitted to finish his service before the mechanism is turned over to the higher-priority customer. Preemptive systems are further classified as *repeat* or *resume*, depending on whether a preempted customer does or does not lose the service already acquired before preemption.

There are other servicing systems which might be considered to be examples of priorities. The student wishing to pursue the question of priorities, using essentially the techniques developed here, is referred to the books of Jaiswal or Cohen.[‡]

[†] It is customary to assign the number 1 to the highest priority class, 2 to the second highest, etc. This can lead to curious locution: for class r is "higher " than class $r+1$.

[‡] JAISWAL, N. K. (1969), *Priority Queues*, Academic Press, New York; COHEN J. W. (1969), *The Single Server Queue*, American Elsevier, New York.

The present section contains a typical result, the mean queueing time for the priority classes of an $M/G/1$ queue. The calculation of the exact probabilities is not really difficult, but in many realistic queueing situations, information on expected values is all that is needed, and this is usually found by a direct argument, such as the one given here.

Consider a queue with priority classes $1, 2, \ldots, N$, with arrival rates for the various classes $\lambda_1, \lambda_2, \ldots, \lambda_N$, service time distributions $b_1(x)$, $b_2(x), \ldots, b_N(x)$, mean values $1/\mu_1, 1/\mu_2, \ldots, 1/\mu_N$, and queueing times $V_1, V_2, \ldots, V_N$. The problem is to express $E(V_k)$ in terms of the other quantities. It is clear, first of all, that the queueing time is composed of three parts for a head-of-the-line system, namely, (i) the residual service of the customer already in service, if there is one, (ii) the time to service all customers of equal or higher priority who are already in the queue, and (iii) the time to service all customers of higher priority who arrive before the customer of priority k can enter service.

Let the mean residual service time be denoted by Ω. Then, since there is no contribution to $E(V_k)$ when the service mechanism is idle, the residual service conditional on the customer in service being of priority j is given by

$$\int_0^\infty x\mu_j \left[1 - B_j(x)\right] dx,$$

according to the results of Section 5.10, where B_j is the distribution function corresponding to b_j. Now the probability of an item of priority j being in service is the product of the arrival rate λ_j and service rate $1/\mu_j$, which will be denoted by ρ_j. Therefore

$$\Omega = \sum_{j=1}^N \rho_j \int_0^\infty x\mu_j \left[1 - B_j(x)\right] dx = \frac{1}{2} \sum_{j=1}^N \lambda_j \int_0^\infty x^2 b_j(x) \, dx. \qquad (77)$$

It is worthwhile at this stage to consider the distributions and parameters as they relate to the queue as a whole, disregarding the priorities. It is first obvious that the total arrival rate is $\lambda = \lambda_1 + \lambda_2 + \cdots + \lambda_N$. Furthermore, the service time distribution is given by

$$b(x) = \frac{1}{\lambda} \sum_{j=1}^N \lambda_j b_j(x). \qquad (78)$$

Hence the mean service time is

$$\frac{1}{\mu} = \frac{1}{\lambda} \sum_{j=1}^{N} \rho_j,$$

and Ω can be written in the form

$$\frac{1}{2} \int_0^\infty x^2 b(x) \, dx. \tag{79}$$

Next, consider the second component of $E(V_k)$, the service times of all customers in queue of priorities $1, 2, \ldots, k$. The expected number of customers of priority j queueing, by Section 6.8, is $\lambda_j E(V_j)$, and the expected service time of such a customer is $1/\mu_j$. Therefore the contribution of these customers to $E(V_k)$ is

$$\sum_{j=1}^{k} \rho_j E(V_j).$$

Finally, the expected number of arrivals of priority j higher than k during the queueing time of a customer of priority k is $\lambda_j E(V_k)$, and the time to service such an arrival is $1/\mu_j$. Thus the contribution of category $j < k$ to $E(V_k)$ is

$$E(V_k) \sum_{j=1}^{k-1} \rho_j.$$

Combining these terms gives the recurrence equation

$$E(V_k) = E(V_k)(\rho_1 + \cdots + \rho_{k-1}) + \rho_1 E(V_1) + \cdots + \rho_k E(V_k) + \Omega.$$

Beginning with $k = 1$, this equation can be solved successively to yield

$$E(V_1) = \frac{\Omega}{1 - \rho_1},$$

$$E(V_2) = \frac{\Omega}{(1 - \rho_1 - \rho_2)(1 - \rho_1)},$$

and so forth.

The general formula

$$E(V_k) = \frac{\Omega}{(1-\rho_1-\rho_2-\cdots-\rho_{k-1})(1-\rho_1-\rho_2-\cdots-\rho_k)} \tag{80}$$

can be verified by the student (using, for example, mathematical induction). It would also be worthwhile to confirm that neither of the denominator factors can become negative. Why does the letter N not enter into this formula?

The analysis of preemptive priorities is simpler with respect to the queueing time, since $\Omega = 0$, but more complex with respect to the duration of service, since the service period may be interrupted repeatedly. These systems are discussed by White and Christie.[†]

6.18. Reverse-Order Service (LIFO)

Another important variation in servicing procedure is to serve the most recent arrival first; this is called LIFO (*last in, first out*). An example would be an inventory where new items are added to the pile and removed as needed from the top. Another would be an in-basket, where jobs are taken from the top for attention. The analysis of LIFO queues is particularly simple in terms of the results of Section 6.14.

As with other service-order permutation problems, the important quantity is waiting time, since the number in the system and the busy period are not affected. To indicate reverse-order service, the superscript R will be used: $\alpha^R(s)$ for the Laplace transform of the queueing time distribution. This transform will now be expressed in terms of $\sigma(s)$, the transform of the continuous busy period, for an $M/G/1$ queue.

First note that if the queue is empty, which has probability $1-\rho$, there is no wait, and so the transform has a term $1-\rho$. Assuming then that the queue is occupied when the customer arrives, the wait will consist of two parts: the residual service time of the customer in service and the continuous busy period induced by that residual service time. Suppose the residual time is fixed at the value τ. Under this condition, the transform of the queueing time consists of the two factors $e^{-s\tau}$ (for the residual service) and $e^{-\lambda\tau+\lambda\tau\sigma(s)}$ (for the continuous busy period induced by τ), according to Theorem 10 of

[†] WHITE, H., and CHRISTIE, L. S. (1958), "Queuing with preemptive priorities or with breakdown," *Operations Research* 6, 79–95.

Section 6.14. The distribution of τ is denoted by $\bar{b}(x)$, so that the unconditional transform of the queueing time can be written, for the case where the service is occupied on arrival,

$$\int_0^\infty e^{-s\tau-\lambda\tau+\lambda\tau\sigma(s)}\bar{b}(\tau)\,d\tau. \tag{81}$$

Evaluating this integral as a Laplace transform $\bar{\beta}$ of $\bar{b}$ and making allowance for the empty server case gives the compact result

$$\alpha^R(s)=1-\rho+\bar{\beta}\left[s+\lambda-\lambda\sigma(s)\right]. \tag{82}$$

By Theorem 4 of Section 5.10,

$$\bar{\beta}\,(s)=\frac{\mu}{s}\left[1-\beta(s)\right],$$

so that

$$\alpha^R(s)=1-\rho+\frac{\lambda\left[1-\beta(s+\lambda-\lambda\sigma(s))\right]}{s+\lambda-\lambda\sigma(s)}.$$

Using Eq. (59) of Section 6.13, the numerator can be simplified, giving the form

$$\alpha^R(s)=1-\rho+\frac{\lambda\left[1-\sigma(s)\right]}{s+\lambda-\lambda\sigma(s)}. \tag{83}$$

The student should show that for the $M/M/1$ case, this reduces to

$$\alpha^R(s)=\tfrac{1}{2}[3-\rho+s/\mu-\Omega/\mu],$$

where

$$\Omega=\left[(\lambda+\mu+s)^2-4\lambda\mu\right]^{1/2}.$$

In comparing this result for a LIFO queue with the corresponding result (Sections 6.8 and 6.9) for a FIFO ("first in first out") queue, it is interesting to see that the means are equal, while the LIFO variance is greater than the corresponding FIFO variance. The proof will be given for the means, and the student with considerable time at his disposal may follow the same general pattern of argument for the variances. It is convenient to define

$$u = \lambda + s - \lambda \sigma(s),$$

so that

$$\sigma(s) = \beta(u)$$

and

$$\frac{du}{ds} = 1 - \lambda \sigma'(s).$$

Then, differentiating the generating function $\alpha^R(s)$ gives

$$\frac{d\alpha^R(s)}{ds} = \frac{\lambda}{u^2} \frac{du}{ds} \left[\beta(u) - u\beta'(u) - 1 \right].$$

This equation must be evaluated at the point $s = 0$. Since $\sigma'(0)$ is minus the continuous busy period mean, given in Section 6.12 as

$$\frac{1}{\mu} \frac{1}{1-\rho},$$

it follows that

$$\frac{du}{ds} \bigg|_{s=0} = \frac{1}{1-\rho}.$$

Therefore the LIFO mean queueing time is given by

$$\frac{-\lambda}{1-\rho} \lim_{s \to 0} \frac{\beta(u) - u\beta'(u) - 1}{u^2},$$

and two applications of L'Hôpital's rule give the simple form

$$\frac{1}{2} \frac{\lambda}{1-\rho} \beta''(0), \tag{84}$$

a form corresponding to the Pollaczek–Khintchine formula for the random variable Y.

Recommendations for Further Study

Cooper (1972) gives an elementary treatment of queueing theory, covering many of the problems discussed here and at about the same level of mathematical sophistication. A number of special topics in the theory can be found in this book.

The excellent small monograph by Cox and Smith (1961) treats many of the topics covered in this chapter with minimal though elegant mathematics. A number of applications are mentioned, with a short chapter on machine interference.

Takács (1962) in one of the earliest books on queueing theory provides a comprehensive but often difficult mathematical treatment of the subject as it appeared 20 years ago. The companion volume (Takács, 1967) displays ingenious combinatorial methods for treating queueing problems, as well as other stochastic processes.

Fortunately, the best book on queueing theory is also one of the most comprehensive, Kleinrock (1975). The author manages to treat a variety of queueing situations in a manner designed to bring out ideas and technique, without going into every possible variant with its own characterization. Nearly every topic covered in this chapter will have its counterpart in the Kleinrock volume, which contains a great deal of additional material.

The small monograph by Newell (1971) attempts to "turn queueing theory around and point it toward the real world" by providing useful approximations for realistic problems.

COOPER, ROBERT B. (1972), *Introduction to Queueing Theory*, Macmillan, New York.

COX, D. R. and SMITH, WALTER L. (1961), *Queues*, John Wiley and Sons, New York.

KLEINROCK, LEONARD (1975), *Queueing Systems, Vol. 1: Theory*, John Wiley and Sons, New York.

NEWELL, GORDON (1971), *Applications of Queueing Theory*, Chapman and Hall, London.

TAKÁCS, LAJOS, (1962), *Introduction to the Theory of Queues*, Oxford University, New York.

TAKÁCS, LAJOS, (1967), *Combinatorial Method in the Theory of Stochastic Processes*, John Wiley and Sons, New York.

6.19. Problems[†]

1. Find the distribution of the number of arrivals in n service periods, both for the $M/M/1$ and $M/D/1$ queues.

[†]Problem 17 is taken from Lindley (1965) (see p. 132), with kind permission of the publisher.

2. Consider a queue in which arrivals and departures are dependent on the number in the system, specifically that λ and μ are replaced by λ_x and μ_x when there are a number x in the system. Show that equilibrium exists only if the series

$$1 + \frac{\lambda_0}{\mu_1} + \frac{\lambda_0 \lambda_1}{\mu_1 \mu_2} + \frac{\lambda_0 \lambda_1 \lambda_2}{\mu_1 \mu_2 \mu_3} + \cdots$$

converges. Consider the special cases: (i) $\lambda_x = \lambda/(x+1)$, $\mu_x = \mu$. This would indicate a queue where new arrivals are discouraged by large numbers in the system, rather similarly to balking. Show that the equilibrium distribution is Poisson. (ii) $\lambda_x = \lambda$, $\mu_x = x\mu$ for $x < N$, $\mu_x = N\mu$, for $x \geq N$. This means that there are N servers. Obtain the equilibrium distribution for the number in the system.

3. In Problem 2, let $\lambda_x = p^x \lambda$, $x \geq 0$, $0 < p < 1$ and $\mu_x = \mu$. (i) Find the equilibrium probability p_x of x customers in the systems in terms of p_0. (ii) Evaluate p_0.

4. In an $M/M/1$ queue, there are only two waiting places available; when these are filled, arrivals are permanently lost. The person in service does not occupy one of the waiting places. Define the states of the system; obtain the equilibrium equations for the state probabilities; find the probability generating function, and by differentiation of this function obtain the mean value of the number in the system.

5. Consider an $M/M/1$ queue with

$$\lambda_x = (n-x)\lambda, \qquad x \leq n,$$

$$\lambda_x = 0, \qquad x > n,$$

$$\mu_x = x\mu.$$

(i) Find explicitly the equilibrium probabilities for the number in the system in terms of λ, μ, and n. (ii) Show that the expected number in the system is $n/(1+1/\rho)$.

6. For an $M/M/n$ queue, find (i) the expected number queueing, (ii) the probability that all servers are occupied, (iii) the probability that there will be someone queueing, (iv) the mean queueing time for those who queue, (v) the mean queueing time for all arrivals, and (vi) the probability that x servers are busy.

7. In an $M/G/\infty$ queue, the service is empty at time zero. Show that the probability of x departures in $[0, t]$ is Poisson with parameter

$$\lambda \int_0^t B(u)\, du.$$

8. Invert $\alpha(s)$ for the $M/D/1$ case to obtain

$$A(t)=(1-\rho)e^{\lambda t}\sum_{j=0}^{[\mu t]}\frac{(\lambda t-j\rho)^{j}}{j!}e^{-j\rho},$$

where $[x]$ denotes the largest integer $\le x$.

9. Find the interdeparture Laplace transform for the $M/G/1$ queue, and verify that it is consistent with the $M/M/1$ result given in Section 6.4.

10. Use the following example to show that the equilibrium distribution of an imbedded process may not agree with the continuous time equilibrium distribution if the points are not regeneration points. In an $M/M/1$ queue, consider instants at which a transition takes place. Then the probability of a jump up is $\lambda/(\lambda+\mu)$ and that of a jump down is $\mu/(\lambda+\mu)$. Write down the transition matrix and show that the equilibrium distribution is

$$p_0=\tfrac{1}{2}(1-\rho),$$

$$p_x=\tfrac{1}{2}(1-\rho)(1+\rho)\rho^{x-1},\qquad x=1,2,3,\dots.$$

11. Use the following outline to given an alternate proof for the probability generating function result of Section 6.6. Let X_n be the number in the system at the end of the nth service, and let N_n be the number of arrivals in the nth service. Show that $X_{n+1}=X_n-\Delta(X_n)+N_{n+1}$, where $\Delta(x)=1, x>0$ and $\Delta(x)=0$, $x\le0$. Form

$$\phi_n(s)=E(s^{X_{n+1}}),$$

and using the independence of X_n and N_n, express this in terms of $\phi_n(s)$ and $\theta(s)$. Pass to the limit as $n\to\infty$.

12. Show that the equilibrium mean number in the $M/M/n$ system is

$$\rho+\frac{n\rho p_n}{(n-\rho)^2}.$$

13. Consider an $M/G/1$ system in which no queue is allowed to form. When the server is occupied, incoming customers are lost. There are two counters for the server, and when both are empty, customers are sent to counter A with probability p and to counter B with probability $1-p$. At each counter the service is negative exponential with parameters $2\mu p$ and $2\mu(1-p)$, respectively. (i) Solve for the equilibrium probability of an empty system. (ii) Find the probability that counter A is occupied. (iii) Find the probability that both counters are occupied.

14. Consider an $M/E_k/1$ system in which no queue is allowed to form. Service is k phases in series, each with negative exponential parameter μk. (i) Find the distribution of X, the number of stages of service left in the system, and (ii) the probability of a busy system.

15. For $\rho > 1$, find the probability of a busy period of finite duration by using the results on the probability of extinction in a branching process.

16. Show that the density function for the length of a single service time, if that service time constitutes a busy period by itself, is

$$\frac{e^{-\lambda x} b(x)}{\beta(\lambda)}.$$

17. You are in a $G/M/1$ queue. Show that the probability that you will have to queue for more than three times as long as the person in front of you is $\frac{1}{4}$.

18. In a $G/M/\infty$ queue, let Z_n be the number of busy servers at the instant of the nth arrival. Show that

$$P(Z_{n+1} = y \mid Z_n = x) = \left(\frac{x+1}{y}\right) \int_0^\infty e^{-\mu y u} (1 - e^{-\mu u})^{x-y+1} \, d(u) \, du.$$

19. In a $D/M/1$ queue with $\rho > 1$ and one person in service and no one queueing, show that the probability that the server will ever again be idle is $1/\rho$.

20. In a $G/D/1$ queue, show that

$$p_x = q_x p_0 + q_x p_1 + q_{x-1} p_2 + \cdots + q_0 p_{x+1}$$

and therefore that

$$\phi(s) = \frac{\theta(s)(1-s)}{\theta(s) - s} \phi(0).$$

Show that the average number of customers queueing is

$$\frac{\frac{1}{2}\theta''(1)}{1 - \theta'(1)}.$$

21. Work out Section 6.15 for E_k arrivals as far as you can. For E_2, express the mean $E(X)$ in terms of ρ.

22. Consider a queue in which arrivals occur in a Poisson process of groups, with g_n being the probability of a group of size n, $n = 1, 2, 3, \ldots$, and probability generating function $g(s) = \Sigma g_j s^j$. (i) Show that the generating function for the number of customers arriving in time t is given by $\exp[\lambda t g(s) - \lambda t]$. (ii) Show

that $\theta(s)=\beta[\lambda-\lambda g(s)]$. Note: in this problem it is assumed that the arrival, service, and group-size variables are independent.

23. Consider an $M/M/1$ system in which customers are impatient. Specifically, upon arrival, customers estimate their queueing time t and then join the queue with probability e^{-kt} and go away with probability $1-e^{-kt}$, where $k>0$ is a constant. They estimate $t=x/\mu$ where the arrival finds x in the system. (i) In terms of p_0, find the equilibrium probabilities p_x of finding x in the system. (ii) Give an expression for p_0. (iii) Under what conditions will equilibrium hold? (iv) For $k\to\infty$, find p_n explicitly. (v) Relate this model to balking.

24. Let $X(t)$ and $Y(t)$ be independent, Poisson random variables with means λt and μt, respectively. Show that the generating function $z(s)$ for the difference $Z(t)=X(t)-Y(t)$ is given by

$$z(s)=\exp\{\lambda t(s-1)+\mu t[(1-s)/s]\}.$$

Apply this result to the following problem. Travelers and taxis arrive independently at a service point at random rates λ and μ, respectively. Let $Z(t)$ denote the queue size at time t, a negative value representing a line of taxis and a positive value representing a queue of travelers. Beginning with $Z(0)=0$, show that the distribution of $Z(t)$ is given by the difference between independent Poisson variables of means λt and μt.

25. A process has two stages of service. Customers arrive in a Poisson process and are served first by a stage with negative exponential service with parameter μ and then by a stage with negative exponential service with parameter v. Let $X_1(t)$ and $X_2(t)$ denote the numbers in the stages at time t. Show that the equilibrium distribution for X_1 and X_2 consists of two independent geometric distributions.

26. Work Problem 25 with the following modified assumptions. The customer enters the system only if the first server is free, and on completing the first service, enters the second service only if that server is free. Customers not entering either service are permanently lost to the system.

27. Consider an $M/M/1$ queue in which the server remains idle until the queue size reaches n. Then the service of these n customers takes place, and also all other arrivals during this period. When at last the queue is empty, and the server idle, the process is repeated.

28. Consider an assembly line moving with uniform speed with items for service spaced along it. A single server moves with the line while serving, and against it with infinite velocity while transferring to the next item. The line has a *barrier* at which service must be broken off and the server is *absorbed*. A server with negative-exponential (λ) service time starts service on an item when it is time T from the barrier. The spacings between items form a renewal process with distribution function $B(x)$. Show that the probability generating function for

the number of items served before absorption satisfies the integral equation

$$\phi(s,T)=\exp\left(-\lambda T+\lambda Ts\int_0^\infty \phi(s,u)\,dB(u)\right).$$

Hint: Let $Z(t)$ be the distance to the barrier at time t, so that $Z(0)=T$ and $Z(t)=T+X(t)-t$, where $X(t)$ is the spacing between the first and last item completed in $(0,t)$. Then the time to absorption

$$\inf(t\,|\,T+X(t)-t\le 0)$$

is the continuous busy period of the queue $M/G/1$.

29. Consider a queue which is $M_\lambda/M_\mu/1$ for ordinary items, but which has one priority item which displaces the item in service according to the preemptive repeat scheme. After being served, the priority item returns after an exponentially distributed time with parameter v.

30. Consider a queue with two priority classes, with independent Poisson arrival rates λ_1 and λ_2 (where $\lambda_1+\lambda_2=1$) and with negative exponential service with parameters μ_1 and μ_2. Within classes, the service is FIFO, and the priority is preemptive resume. Let p_{xy} be the equilibrium probability of x priority and y nonpriority customers in the system. Show that

$$\left[\lambda_1+\lambda_2+\mu_1(1-\delta_{x0})+\mu_2(1-\delta_{y0})\delta_{x0}\right]p_{xy}$$

$$=\lambda_1 p_{x-1,y}+\lambda_2 p_{x,y-1}+\mu_1 p_{x+1,y}+\mu_2\delta_{x0}p_{x,y+1},$$

where $\delta_{xy}=0$, $x\ne y$, and $\delta_{xx}=1$, and any symbol with a negative subscript is interpreted as being equal to zero. Show that the mean number of nonpriority customers in the system is

$$\frac{\rho_2}{1-\rho_1-\rho_2}\left(1+\frac{\mu_2\rho_1}{\mu_1(1-\rho_1)}\right),$$

where $\rho_j=\lambda_j/\mu_j$.

31. In an $M/M/1$ queue, suppose there are only a finite number n of possible customers; any customer not in the system at time t has a constant probability $p\Delta t$ of joining the queue during $(t,t+\Delta t)$. Find the equilibrium distribution for the number in the system.

32. Consider a queue with M_λ arrivals, where service is instantaneous but only available at service instants, which are separated by equidistributed intervals, with distribution function $H(x)$, with $\varepsilon=\mathcal{L}H$. Suppose that the maximum number that can be served in a service instant is k, and let Z_j denote the number of customers in the system just before the jth service instant. (i) Show

that

$$Z_{j+1} = Z_j + N_j - k, \qquad Z_j \geq k,$$

$$Z_{j+1} = N_j, \qquad\qquad Z_j < k,$$

where N_j is the number of arrivals in the interval between the jth and $(j+1)$th service instants. (ii) Show that the probability generating function of N_j is $\varepsilon(\lambda(1-s))$, and find the equilibrium probability generating function for Z. (iii) Discuss the applicability of the model to a queue of vehicles at a traffic light.

33. An $M/M/1$ queue *with feedback* has the following defining property: When a customer finishes service, departure occurs with probability p; with probability $1-p$, the customer rejoins the end of the queue. (i) Find the equilibrium probabilities for the number in the system. (ii) Find the probability that a customer is served x times. (iii) Find the expected total time a customer spends in the system.

34. In an $M/M/1$ queue, the service rate is μ only when there are less than three customers in the system. When there are three or more, another server is employed, so that the service rate is increased to 2μ. Formulate and find the equilibrium probabilities.

35. Consider a bulk service system. Whenever the server becomes free, he accepts two customers from the queue into service simultaneously, or, if only one is in the queue, he accepts that one. In either case, the service time for the group (of size 1 or 2) is distributed as $b(x)$ with mean $1/\mu$. Let X_n be the number of customers in the system after the nth service, with equilibrium probability generating function $\phi(s)$, and let N_n be the number of arrivals during the nth service, with equilibrium probability generating function $\theta(s)$. Define $\rho = \lambda/2\mu$. (i) Show that $\theta(s) = \beta(\lambda - \lambda s)$. (ii) Find $E(X)$ in terms of var(X) and $P(X_n = 0)$. (iii) Express $\phi(s)$ in terms of $\beta(s)$, $P(X_n = 0)$, and $P(X_n = 1)$. (iv) Express $P(X_n = 1)$ in terms of $P(X_n = 0)$.

36. Consider an $M/M/\infty$ queue with batch arrivals; the batch size being geometrically distributed with parameter r. Formulate the number in the system as a continuous time Markov process and find the infinitesimal matrix of the process. Find the probability generating function for the equilibrium distribution of the process.

37. Show that the results of Section 6.5, Eqs. (16) and (17), can be obtained as a special case of Section 5.6, Eqs. (21) and (22).

Index

Numbers in italics refer to entire sections.